Student Edition

Eureka Math

Grade K

Modules 5 & 6

Special thanks go to the Gordon A. Cain Center and to the Department of Mathematics at Louisiana State University for their support in the development of *Eureka Math*.

For a free *Eureka Math* Teacher Resource Pack, Parent Tip Sheets, and more please visit www.Eureka.tools

Published by the non-profit Great Minds

Printed in the U.S.A.
This book may be purchased from the publisher at eureka-math.org
10 9 8 7 6 5 4 3

ISBN 978-1-63255-286-0

Name ______________________________ Date ______________

Circle the groups that have 10 ones.

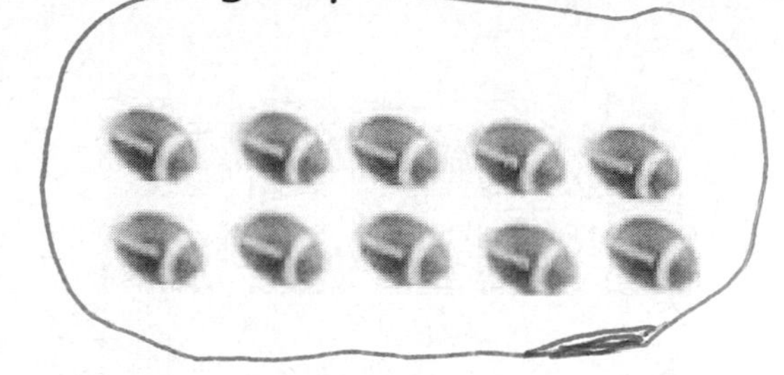

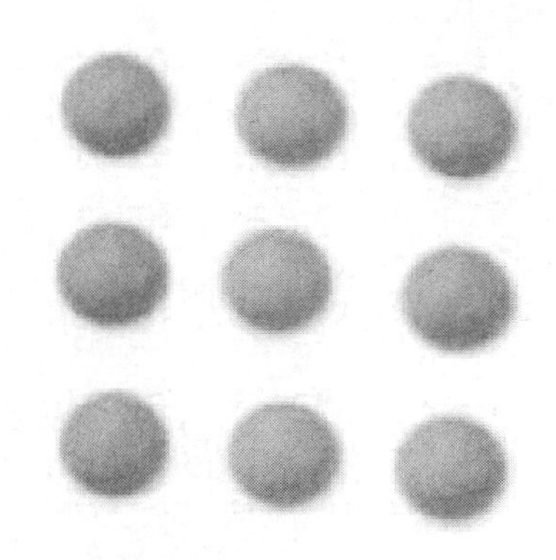

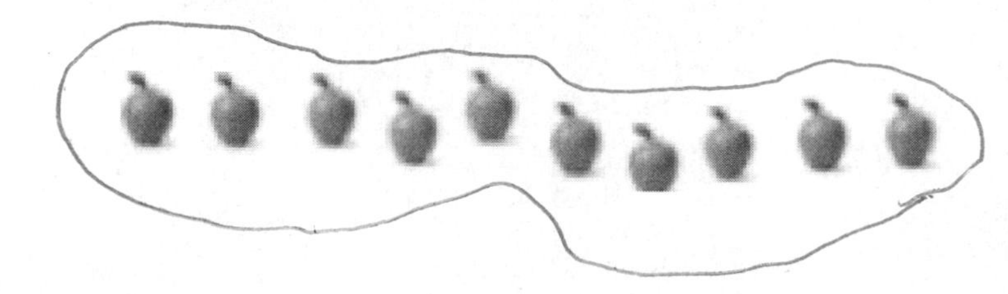

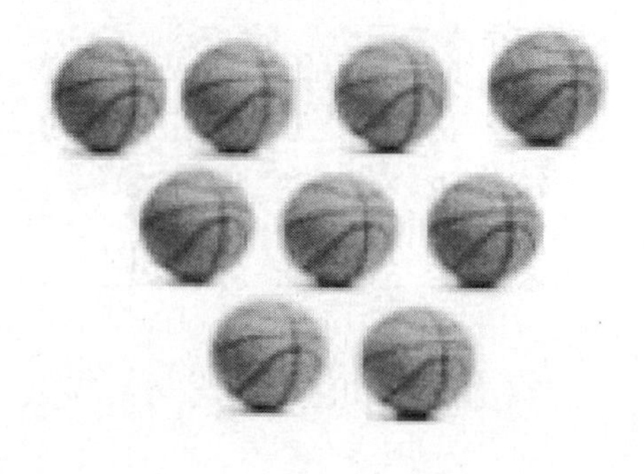

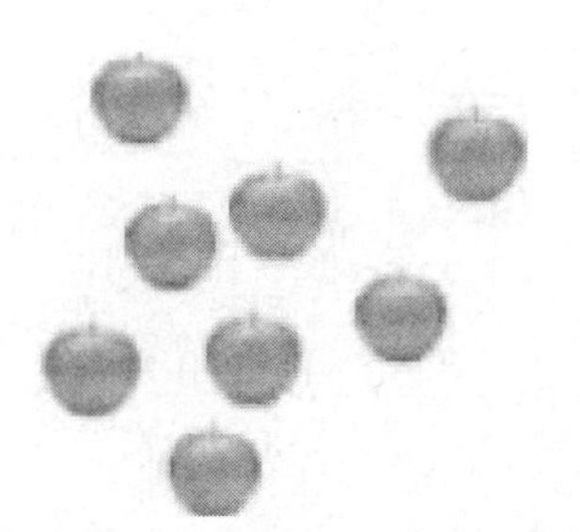

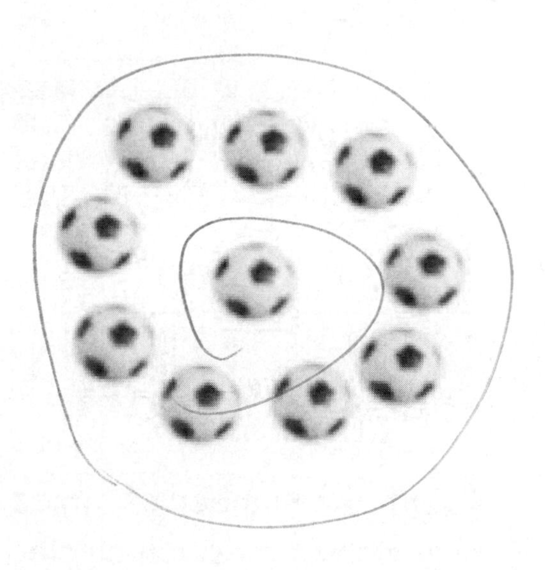

How many times did you count 10 ones?

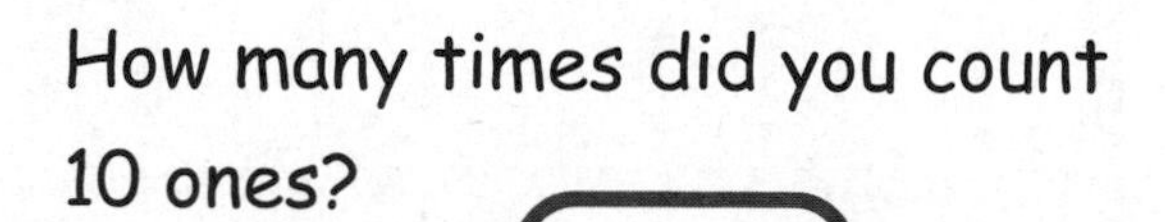

Name ___________________________ Date _______________

Circle 10.

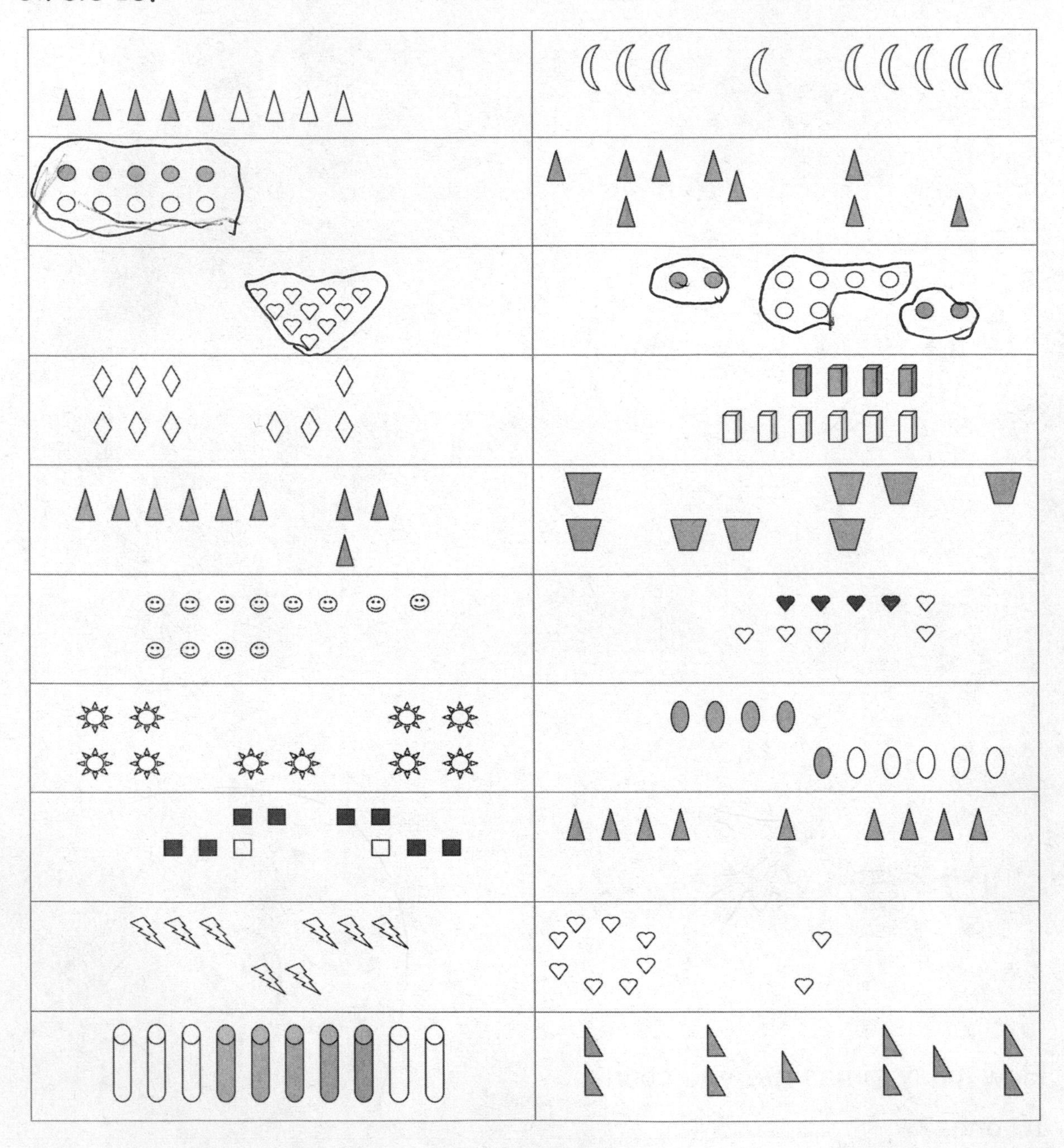

Count the number of times you circled 10 ones. Tell a friend or an adult how many times you circled 10 ones.

Name ______________________________ Date ______________

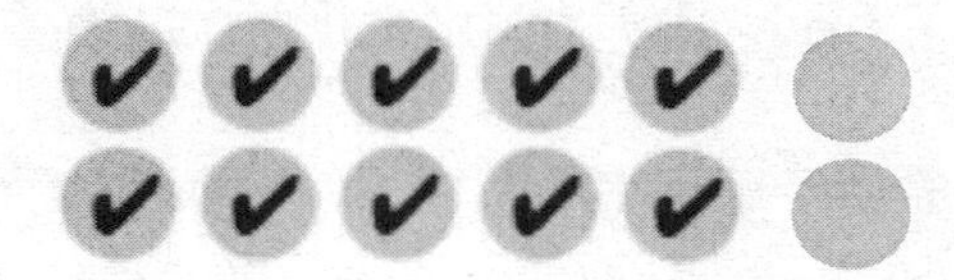

I have 10 ones and 2 ones.

Touch and count 10 things. Put a check over each one as you count 10 things.

I have 10 ones and ___ ones.

I have 10 ones and ___ ones.

I have ___ ones and ___ ones.

I have ___ ones and ___ ones.

Draw pictures to match the words.

I have 10 small circles and 2 small circles:

I have 10 ones and 4 ones:

Name ______________________ Date ____________

10 ones and 3 ones

Draw more to show the number.

10 ones and 2 ones

10 ones and 5 ones

10 ones and 7 ones

10 ones and 4 ones

This page intentionally left blank

Name ______________________________ Date ______________

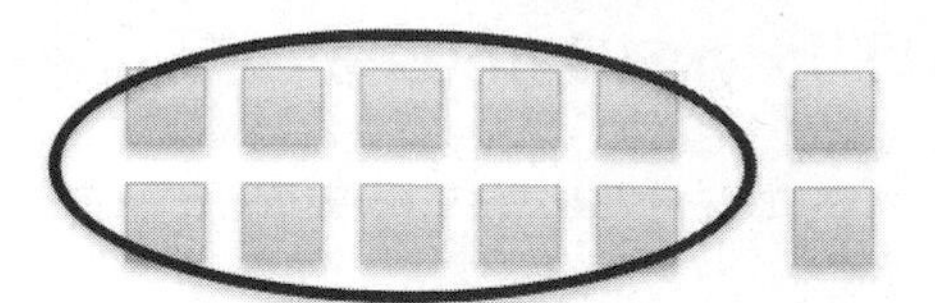

I have 10 ones and 2 ones.

Count and circle 10 things. Tell how many there are in two parts, 10 ones and some more ones.

I have 10 ones and ___ ones.

I have ___ ones and ___ ones.

I have ___ ones and ___ ones.

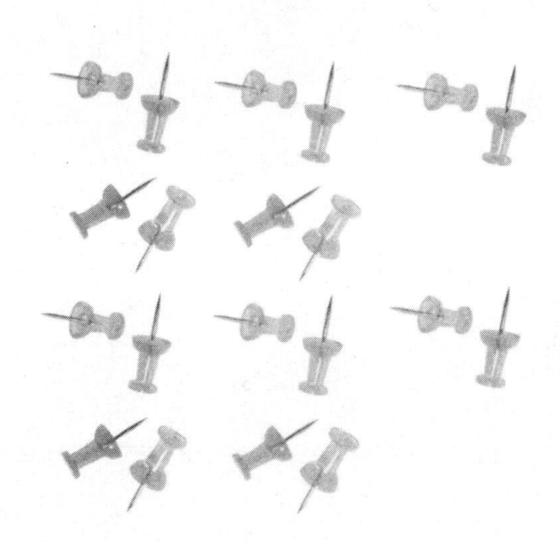

I have ___ ones and ___ ones.

Draw your picture to match the words. Circle 10 ones.

I have 10 ones and 3 ones:

I have 10 ones and 8 ones:

Name ______________________ Date ____________

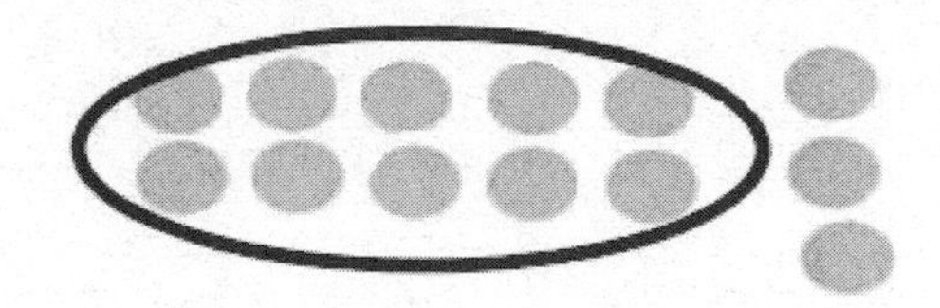

I have 10 ones and 3 ones.

Circle 10 things. Tell how many there are in two parts, 10 ones and some more ones.

I have 10 ones and ___ ones.

I have 10 ones and ___ ones.

I have ___ ones and ___ ones.

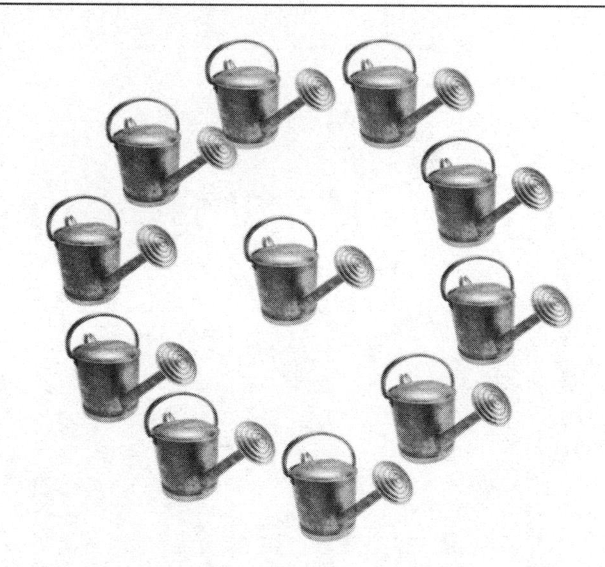

I have ___ ones and ___ ones.

This page intentionally left blank

Name ______________________ Date______________________

Draw 10 ones and some ones. Whisper count as you work the Say Ten Way.

I can make **ten three.**
10 3

I can make **ten seven.**
10 7

I can make **ten two.**
10 2

I can make **ten nine.**
10 9

Name ________________________ Date ______________

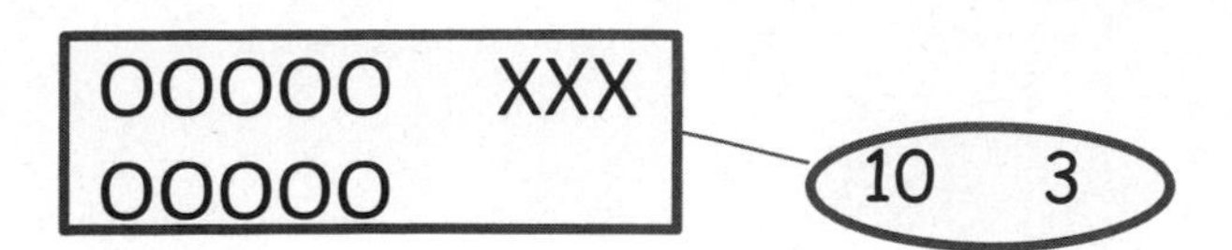

Draw a line to match each picture with the numbers the Say Ten way.

OOOOO X OOOOO	10 1
OOOOO XX OOOOO	10 6
OOOOO XXX OOOOO	10 10
OOOOO XXXXX OOOOO X	10 2
OOOOO XXXXX OOOOO XXXXX	10 3

This page intentionally left blank

Name ____________________ Date ____________

Circle 10 things. Touch and count the Say Ten way. Count your 10 ones first.
Put a check over the loose ones. Draw a line to match the number.

Ten one
10 1

Ten seven
10 7

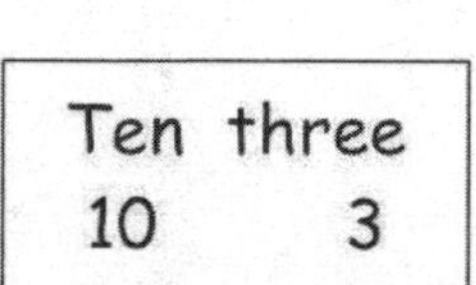

Ten three
10 3

four
10 4

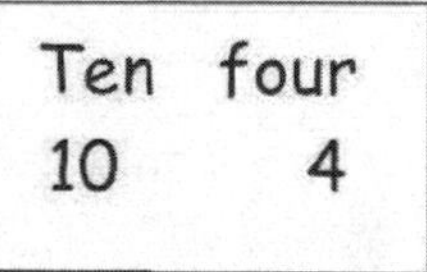

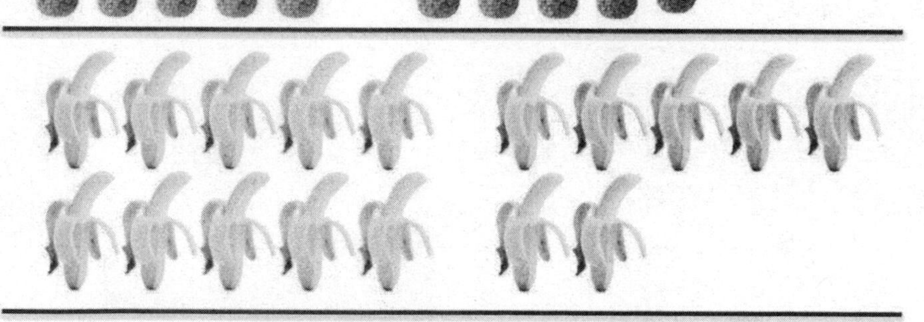

Ten eight
10 8

Lesson 5: Count straws the Say Ten way to 20; make a pile for each ten.

Name ____________________ Date ____________

Write the numbers that go before and after, counting the Say Ten way.

BEFORE	NUMBER	AFTER
10 and 3	10 and 4	10 and 5
and	10 and 2	and
and	10 and 5	and
and	10 and 6	and
and	10 and 1	and
and	10 and 9	and

Name ______________________ Date ______________

Write and draw the number. Use your Hide Zero cards to help you.

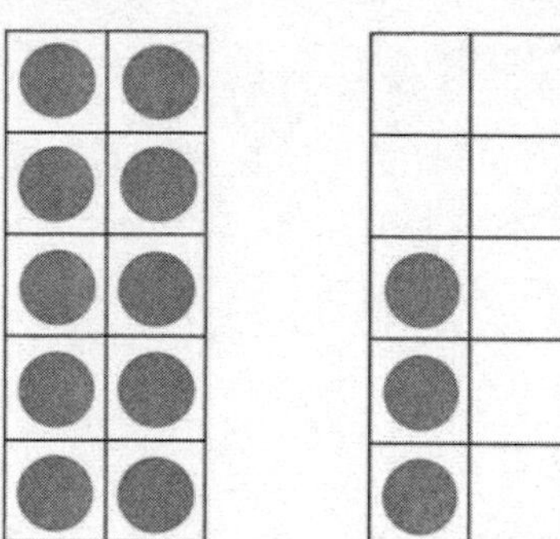

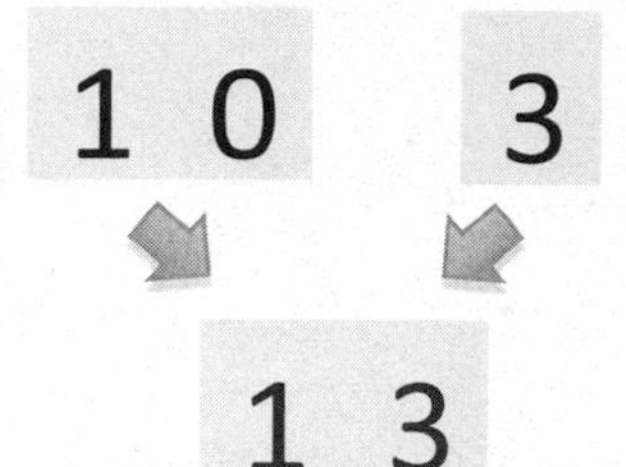

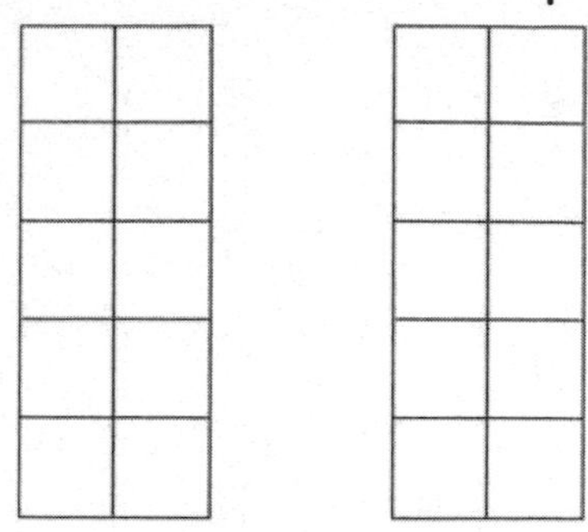

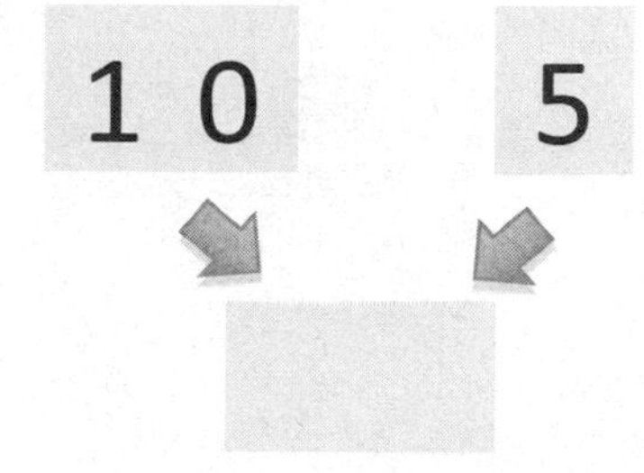

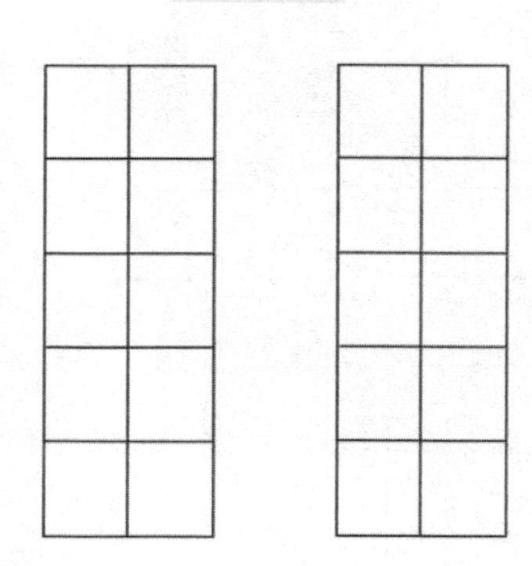

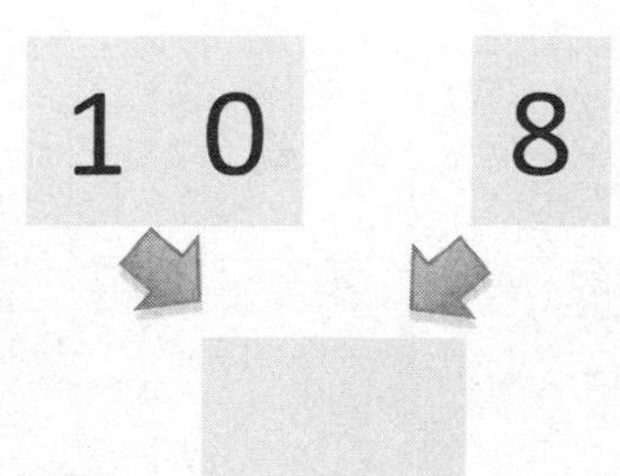

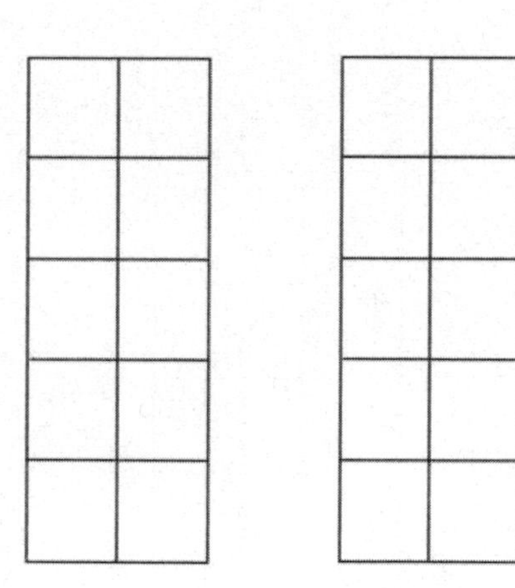

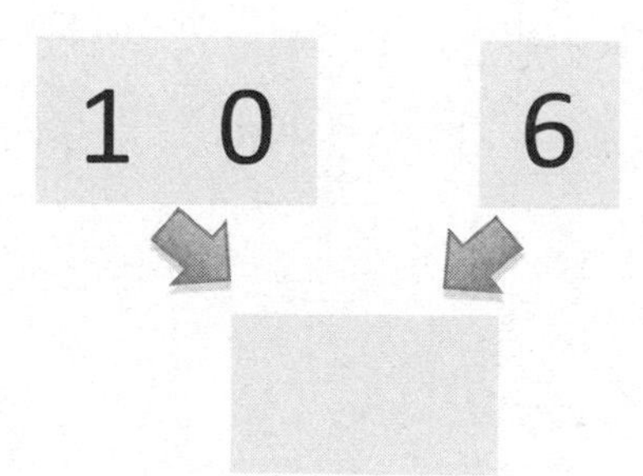

Name ______________________________ Date ______________

Write and draw the number. Use your Hide Zero cards to help you.

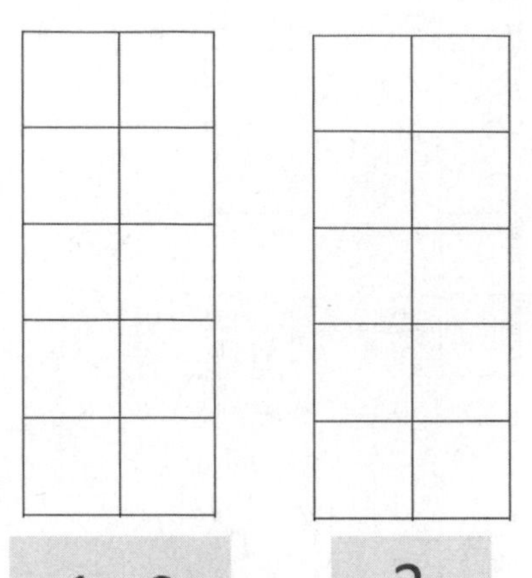

1 0 2

1 0 7

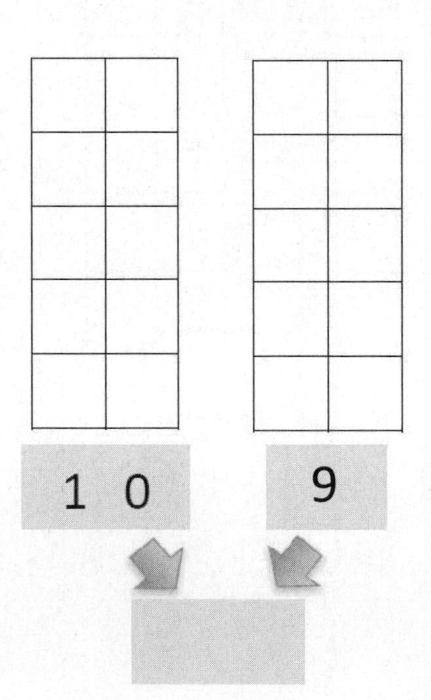

1 0 9

1 0 4

Name ____________________ Date ____________

Look at the Hide Zero cards or the 5-group cards. Use your cards to show the number. Write the number as a number bond.

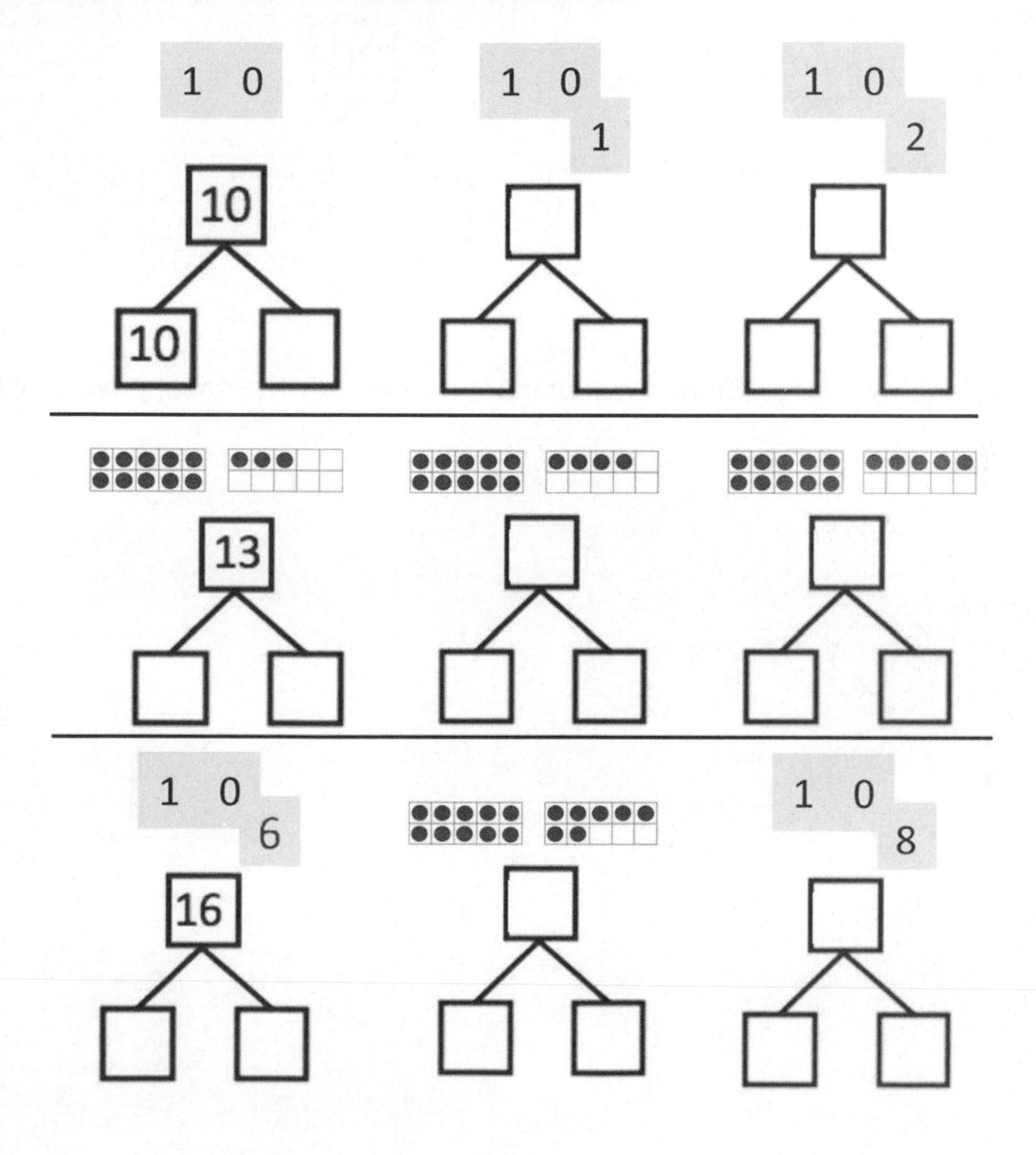

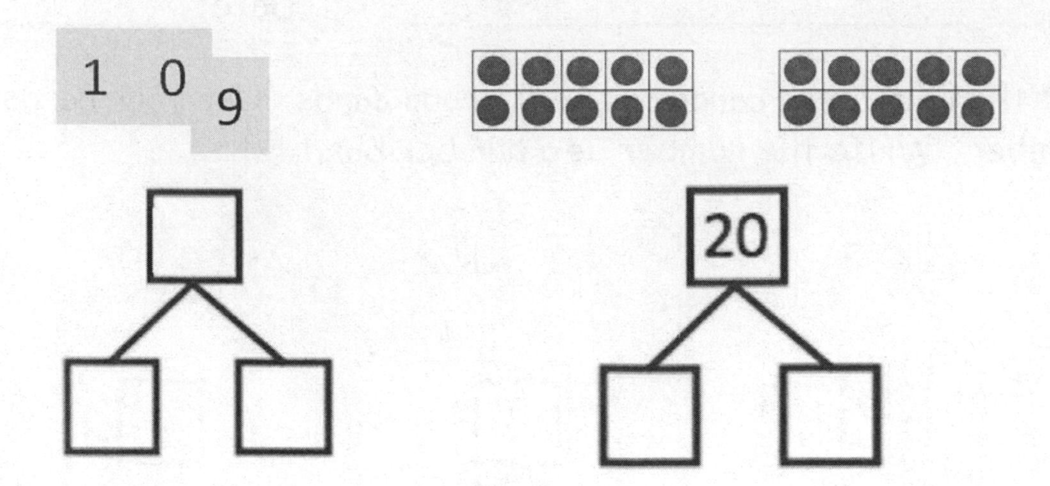

Circle 10 smiley faces. Draw a number bond to match the total number of faces.

Name ______________________ Date ______________

Look at the Hide Zero cards or the 5-group cards. Use your cards to show the number. Write the number as a number bond.

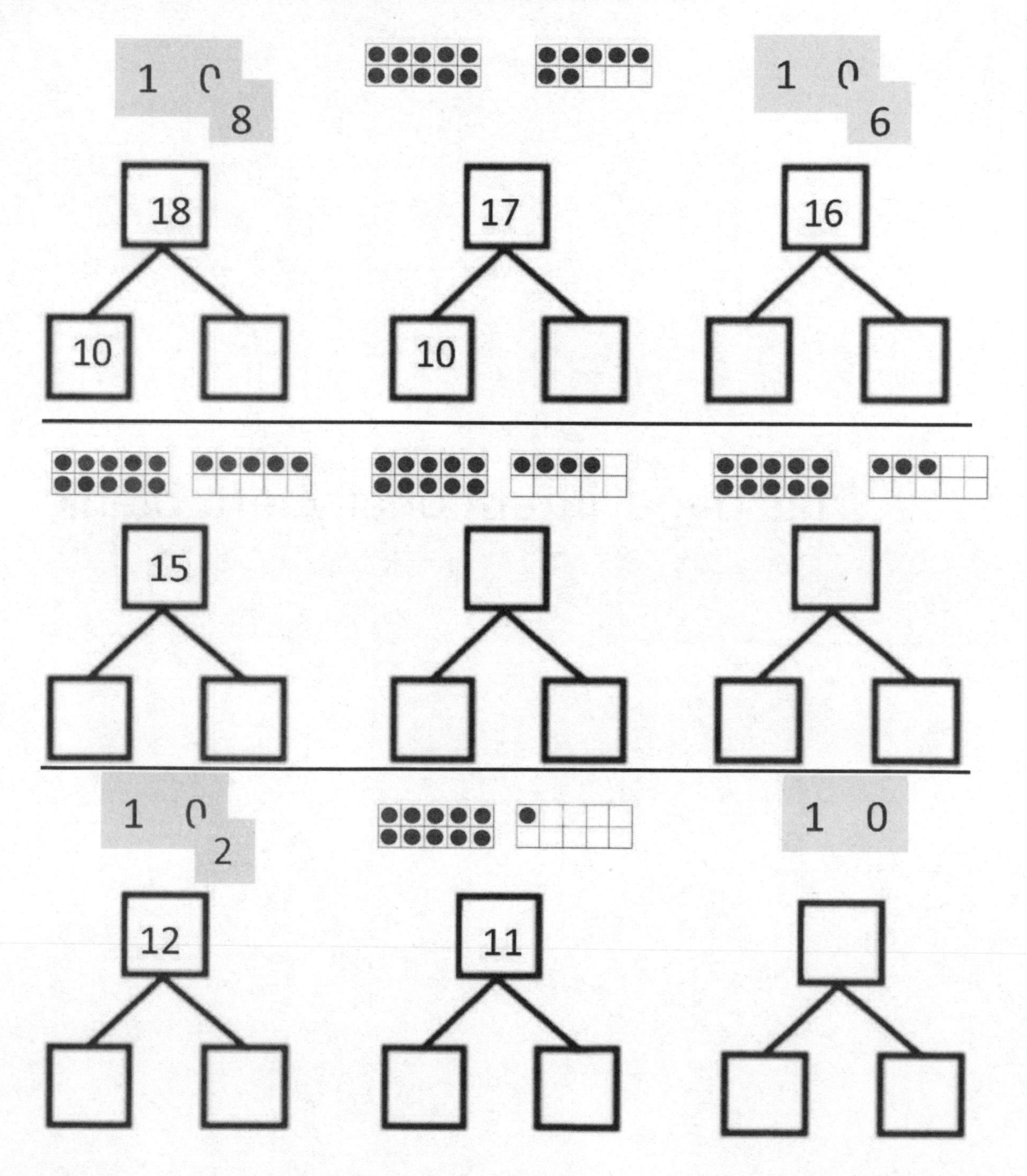

This page intentionally left blank

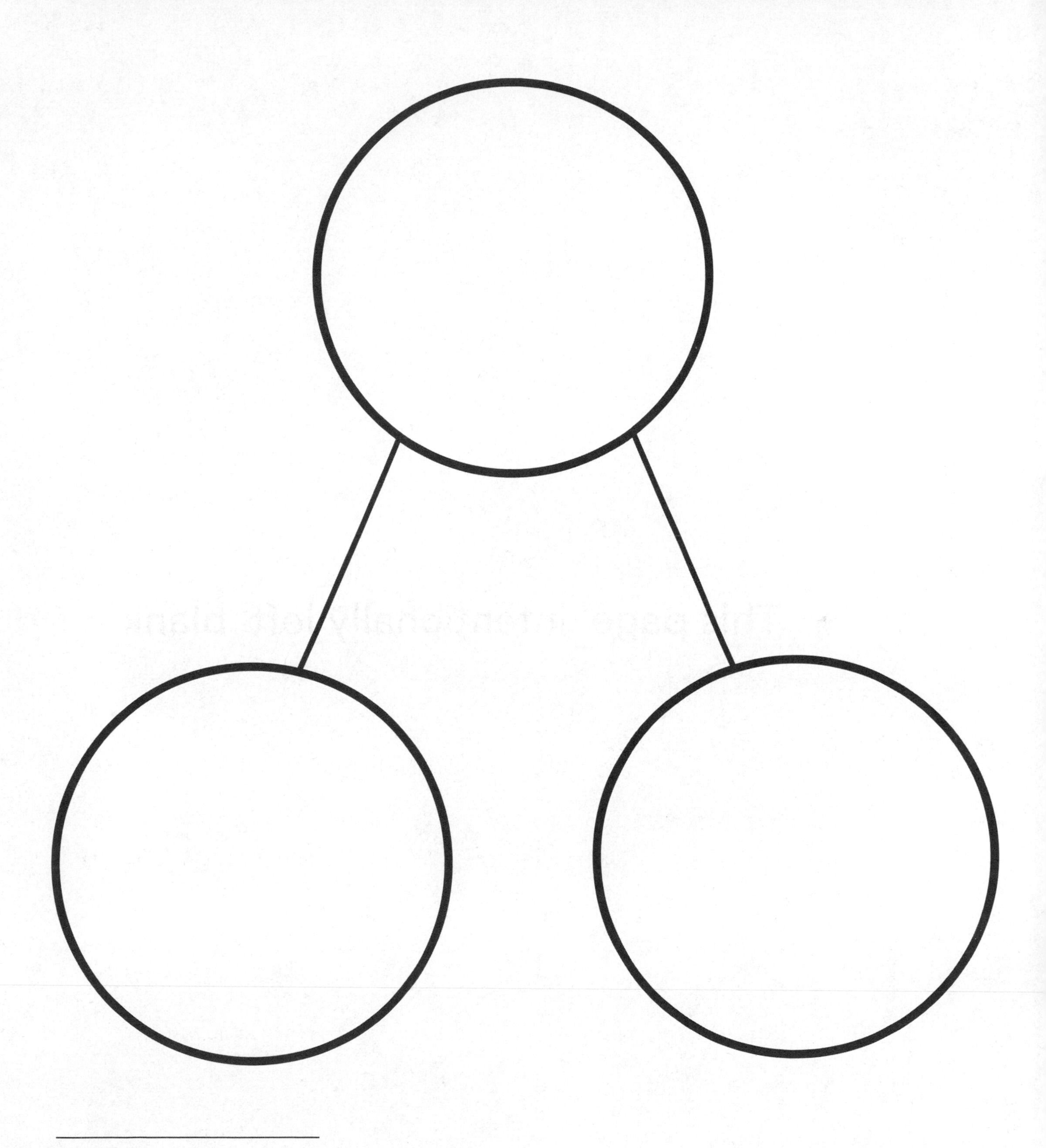

number bond

This page intentionally left blank

Name ______________________________ Date ______________

Use your materials to show each number as 10 ones and some more ones. Use your 5-groups way of drawing. Show each number with your Hide Zero cards. Whisper count as you work.

11	18
15	14

12	17
20	13

Name ______________________________ Date ______________

Use your materials to show each number as 10 ones and some more ones. Use your 5-groups way of drawing.

1 5	1 3
Ten seven	Ten one

1 2	1 6
2 ten	Ten four

Name ____________________ Date ____________

Whisper count as you draw the number. Fill one 10-frame first. Show your numbers with your Hide Zero cards.

12

17

16

13

Draw and circle 10 ones and some more ones to show each number.

20

11

Choose a teen number to draw. Circle 10 ones and some ones to show each number.

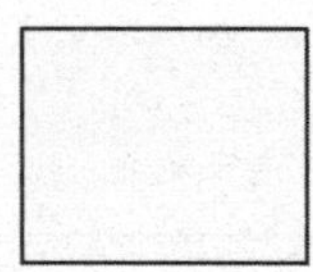

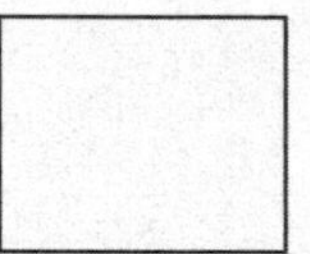

Name ____________________ Date ____________

For each number, make a drawing that shows that many objects.
Circle 10 ones.

11

16

20

19

14

12

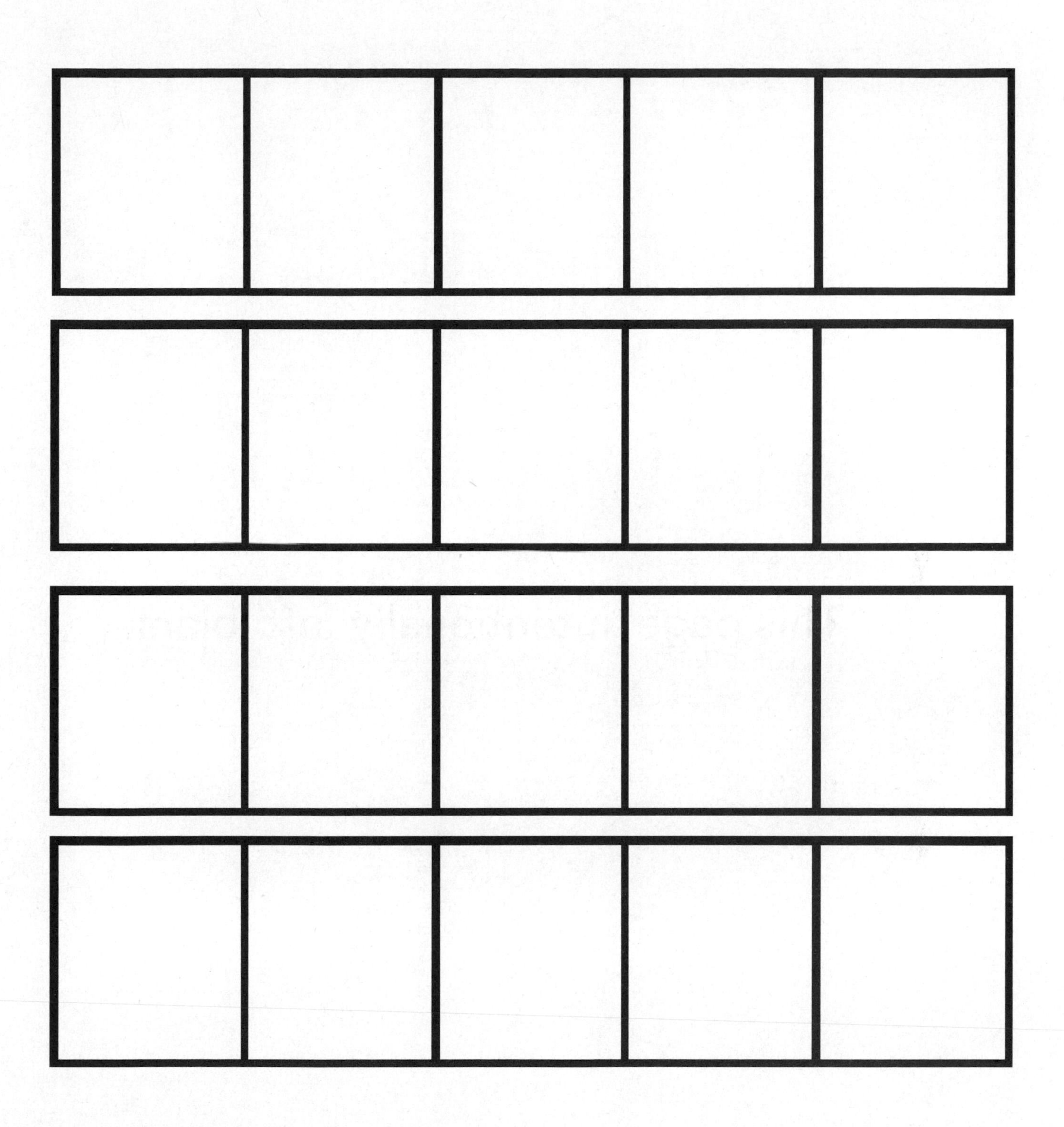

double 10-frame

This page intentionally left blank

Name ______________________ Date ____________

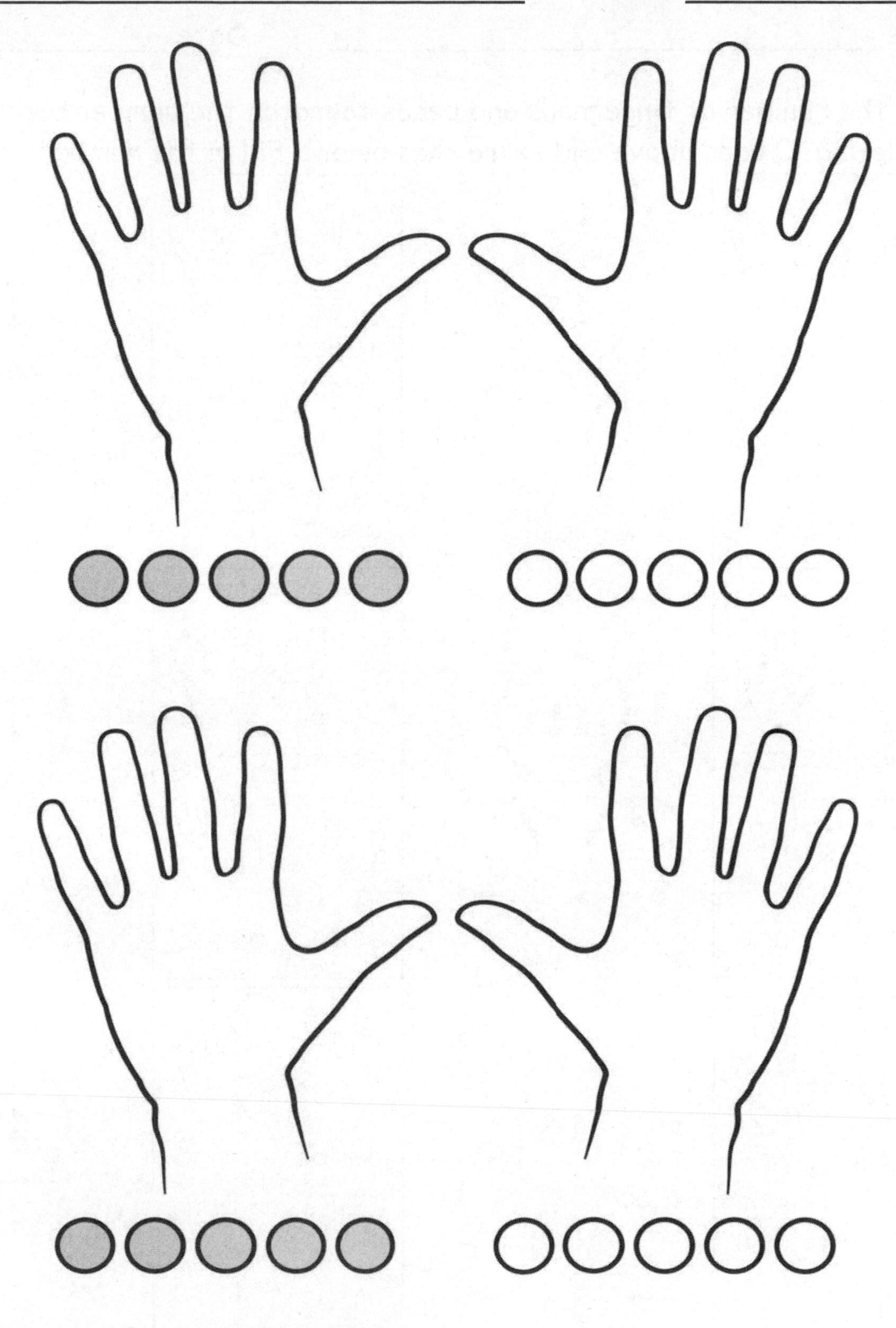

Name ______________________________ Date ______________

Color the number of fingernails and beads to match the number bond. Show by coloring 10 ones above and extra ones below. Fill in the number bonds.

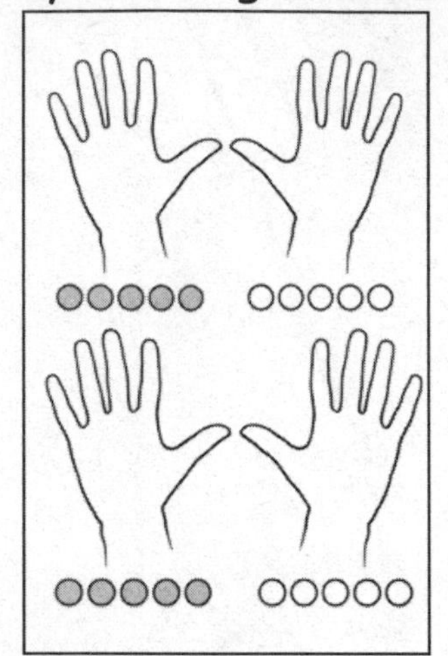

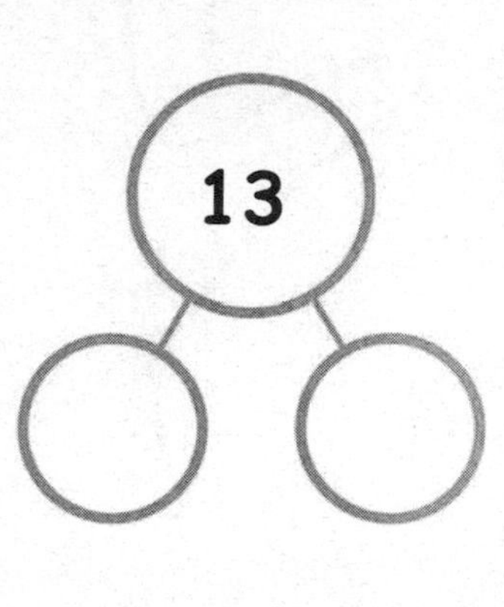

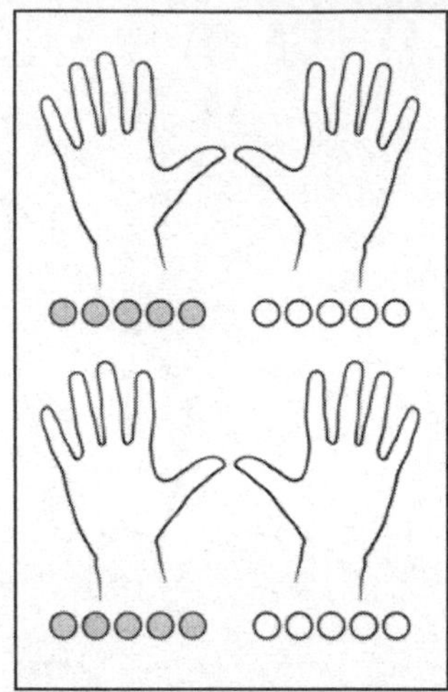

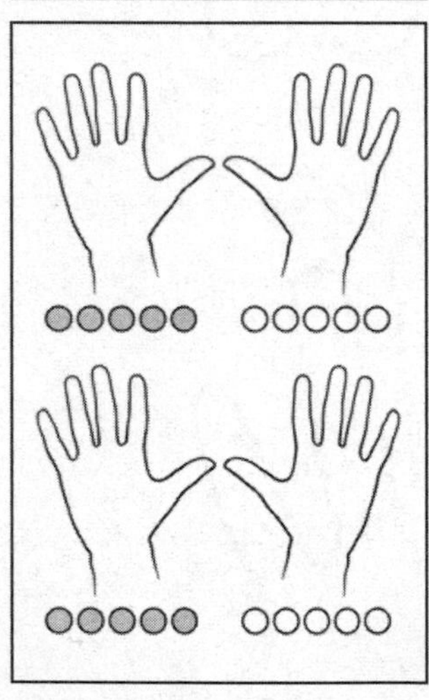

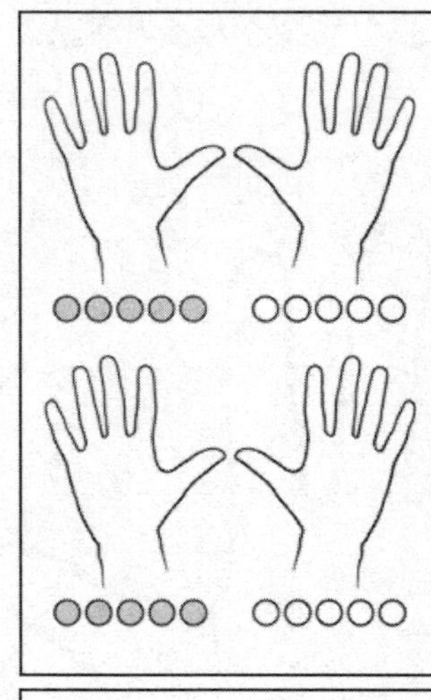

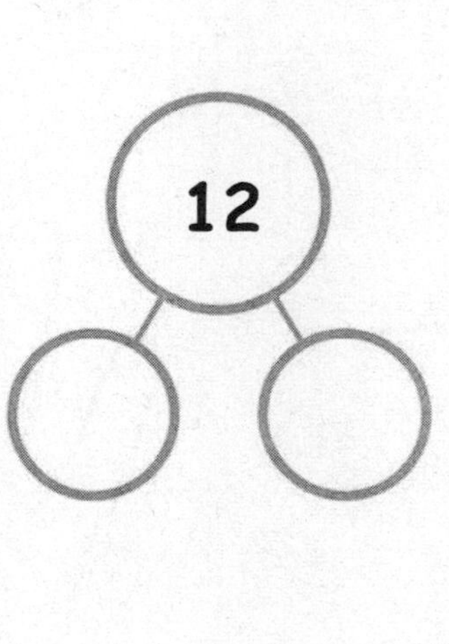

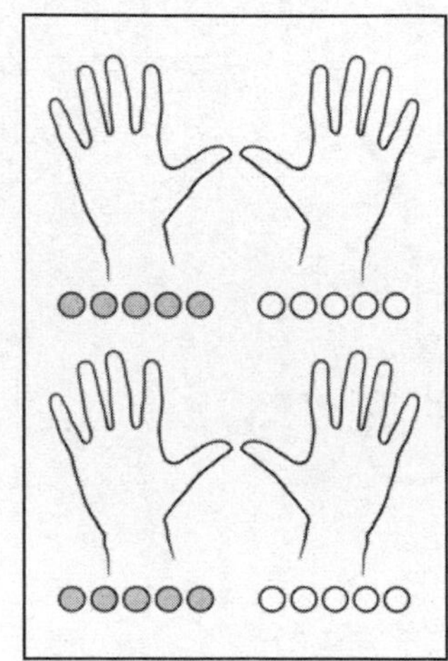

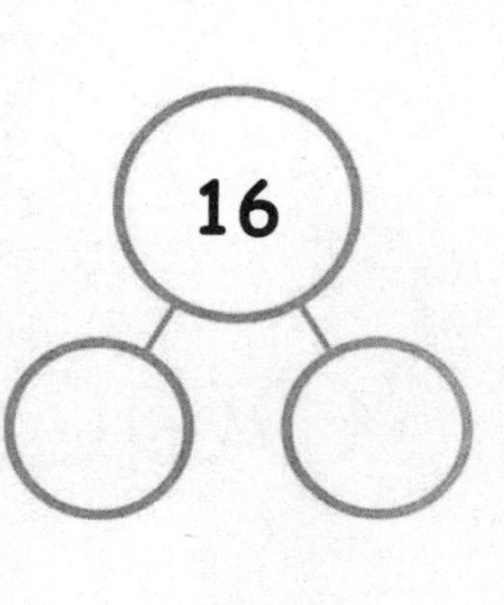

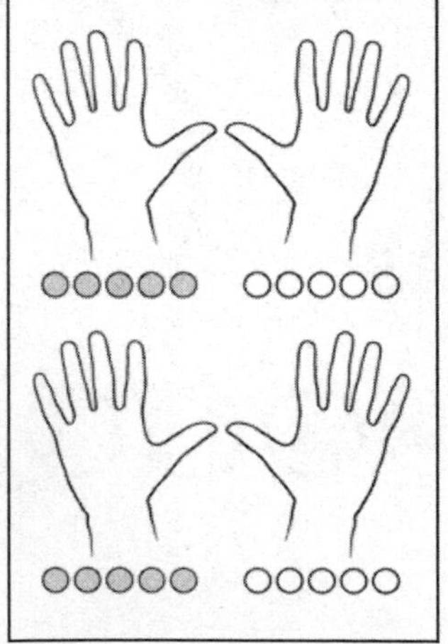

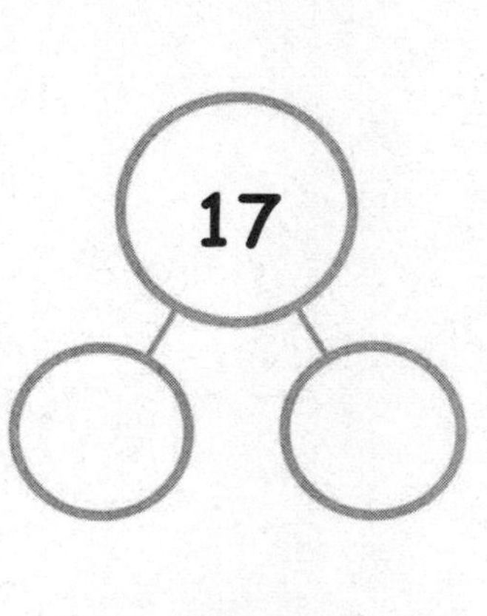

Name ______________________ Date ____________

Count, color and write.

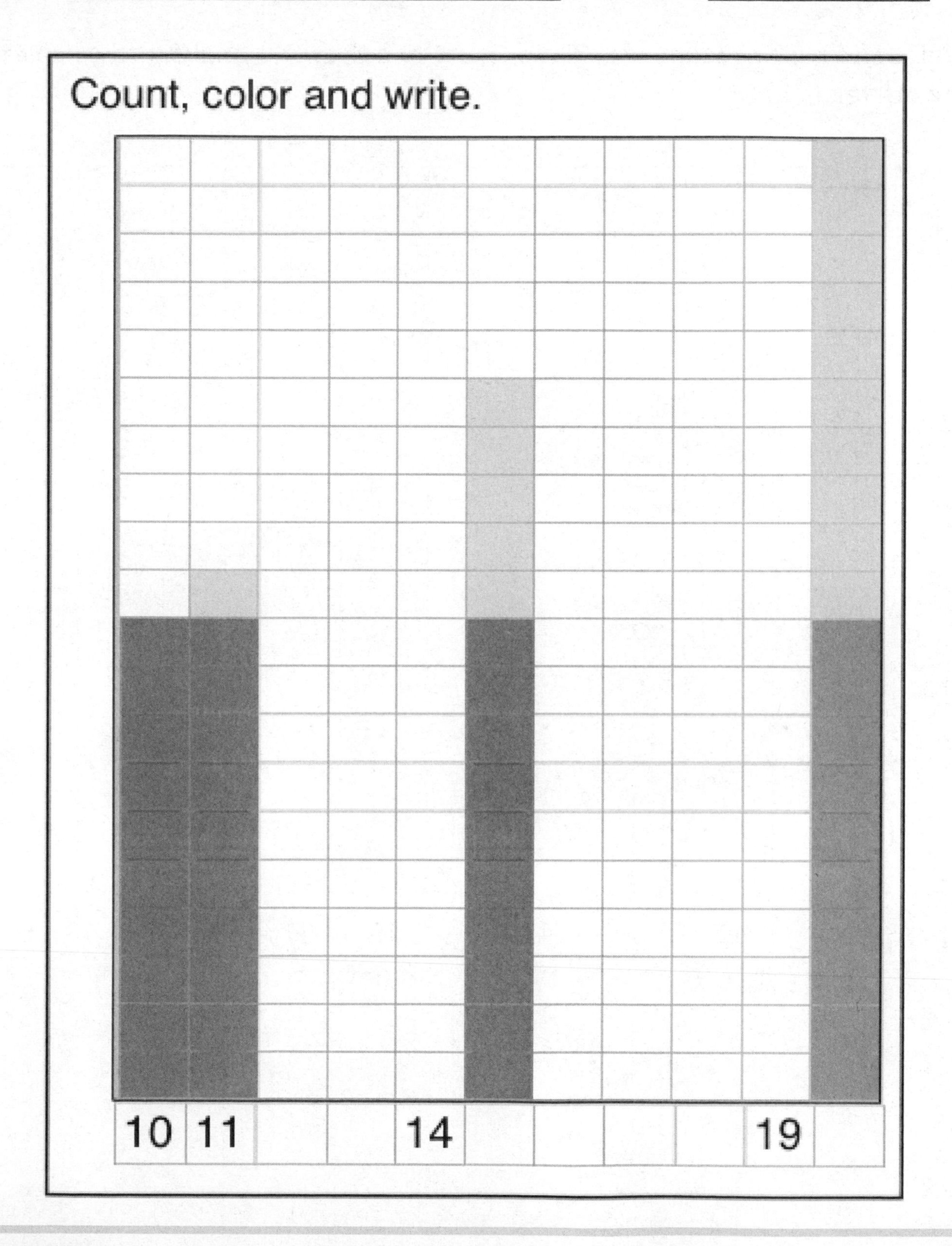

Name ______________________________ Date ______________

Write the missing numbers. Then, count and draw X's and O's to complete the pattern.

O O O O O O O O O O	X O O O O O O O O O O		X X X O O O O O O O O O O		X X X X X O O O O O O O O O O				X X X X X X X X X O O O O O O O O O O	
10		12		14		16	17	18		20

EUREKA MATH™

Name ____________________ Date __________

Count, color and write.

20	19				15	14			11	

Name ______________________________ Date ______________

Write the missing numbers. Then, draw X's and O's to complete the pattern.

X X X X X X X X X X O O O O O O O O O O	X X X X X X X X X O O O O O O O O O O		X X X X X X X O O O O O O O O O O		X X X X X O O O O O O O O O O				X O O O O O O O O O O	
20		18		16		14	13	12		10

Name ______________________________ Date ______________

The ducks found some tasty fish to eat in the boxes!
Count up on the number path.

Write the missing numbers for the boxes that have a duck on top.

______ ______ ______ ______ ______

Write the missing numbers for the boxes that have a duck on top.

______ ______ ______ ______ ______

How many ducks do you count?

_____ _____

In the space below, draw 15 circles in rows.

In the space below, draw 12 squares in rows.

Name ______________________ Date ____________

Count the objects. Draw dots to show the same number on the double 10-frames.

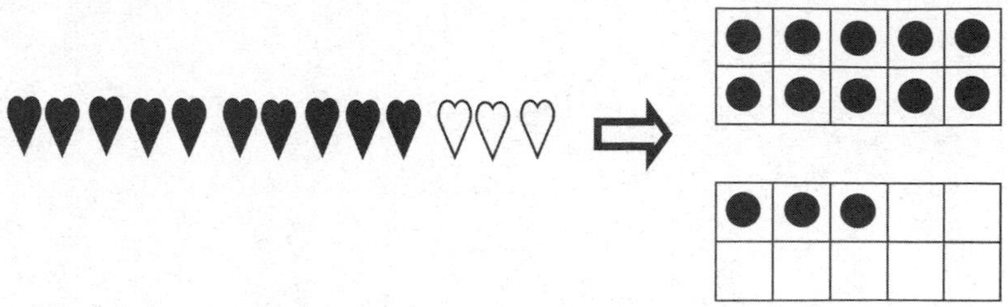

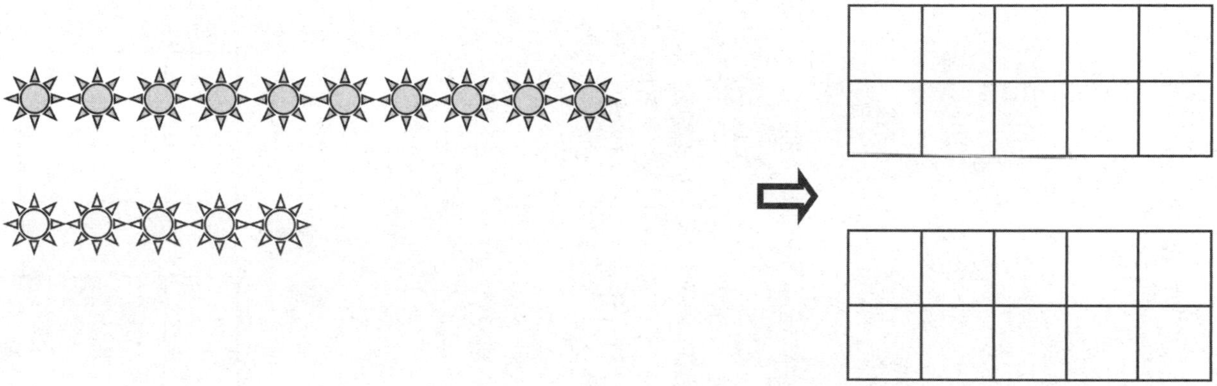

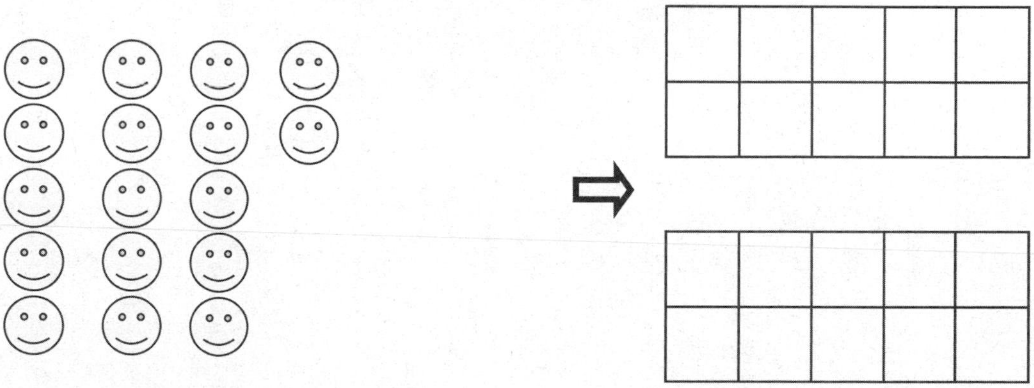

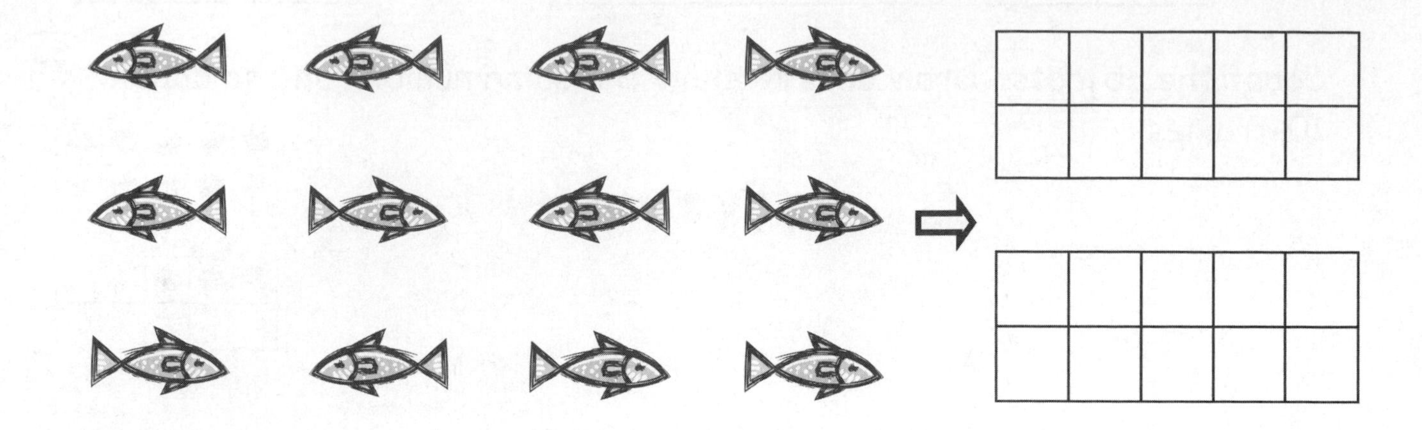

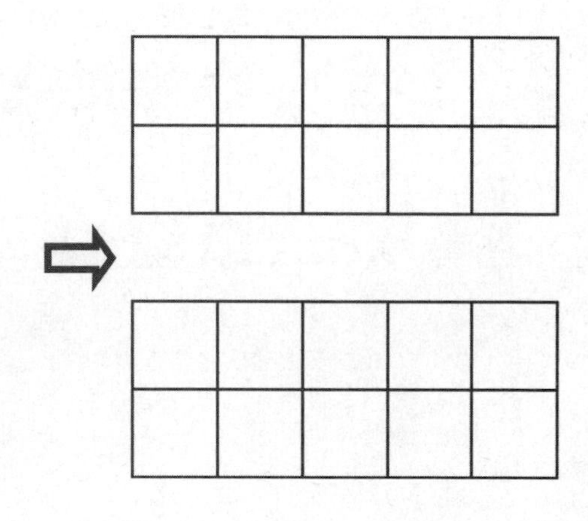

Name ______________________ Date ____________

Whisper count how many objects there are. Write the number.

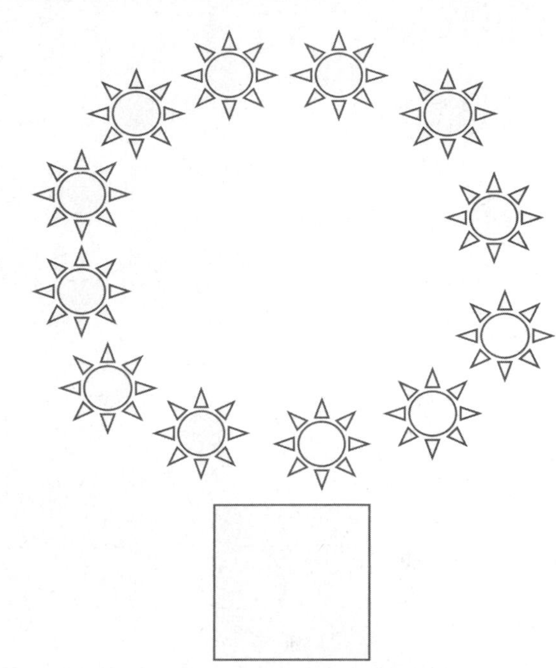

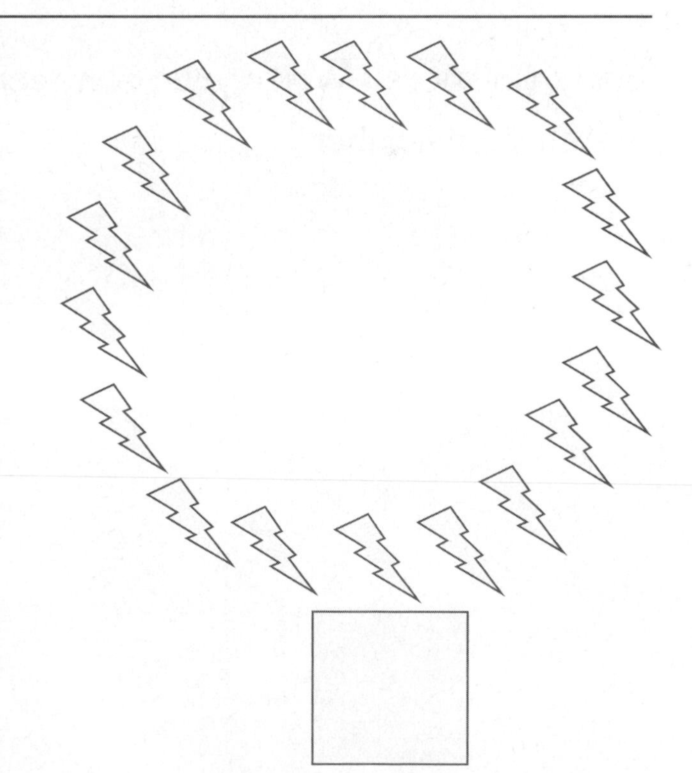

Whisper count and draw in more shapes to match the number.

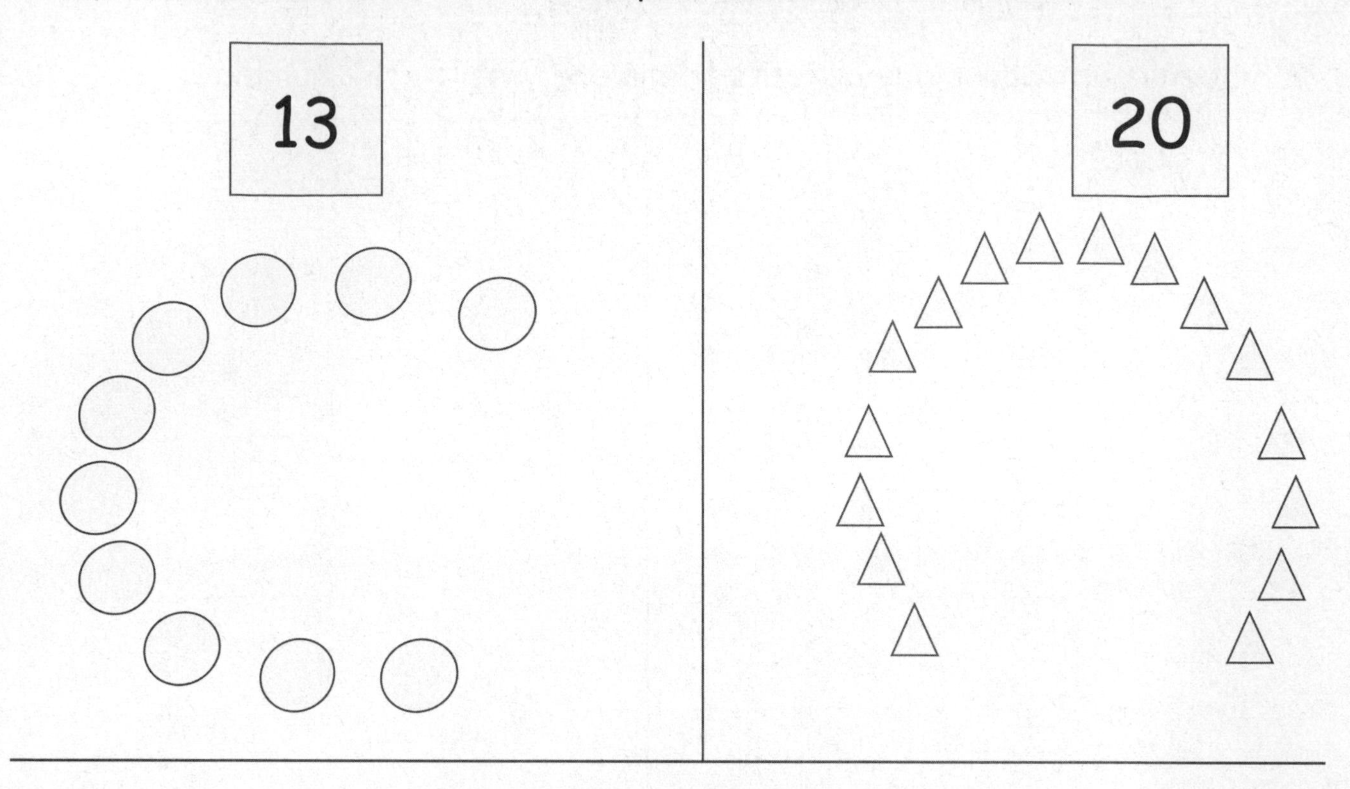

Early finishers: Write your own teen number in the box. Draw a picture to match your number.

Name ____________________________ Date ______________

Count the objects in each group. Write the number in the boxes below the pictures.

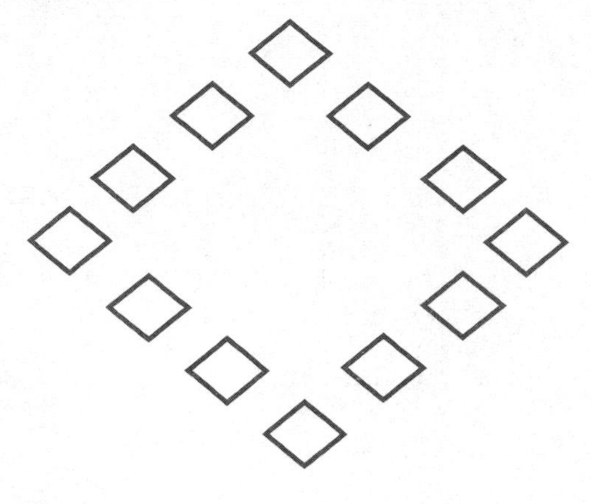

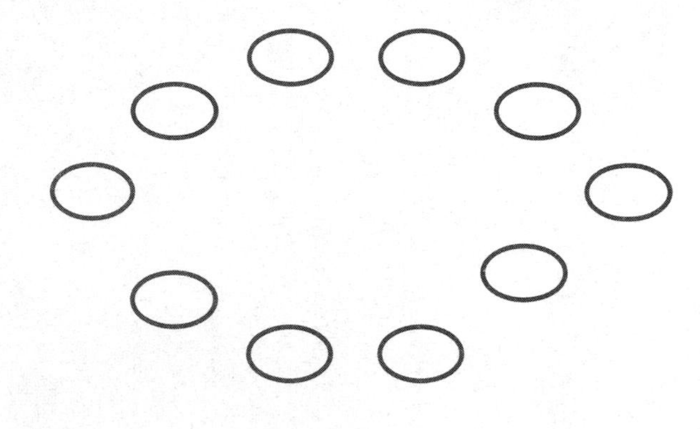

Count and draw in more shapes to match the number.

19

Count the dots. Draw each dot in the 10-frame. Write the number in the box below the 10-frames.

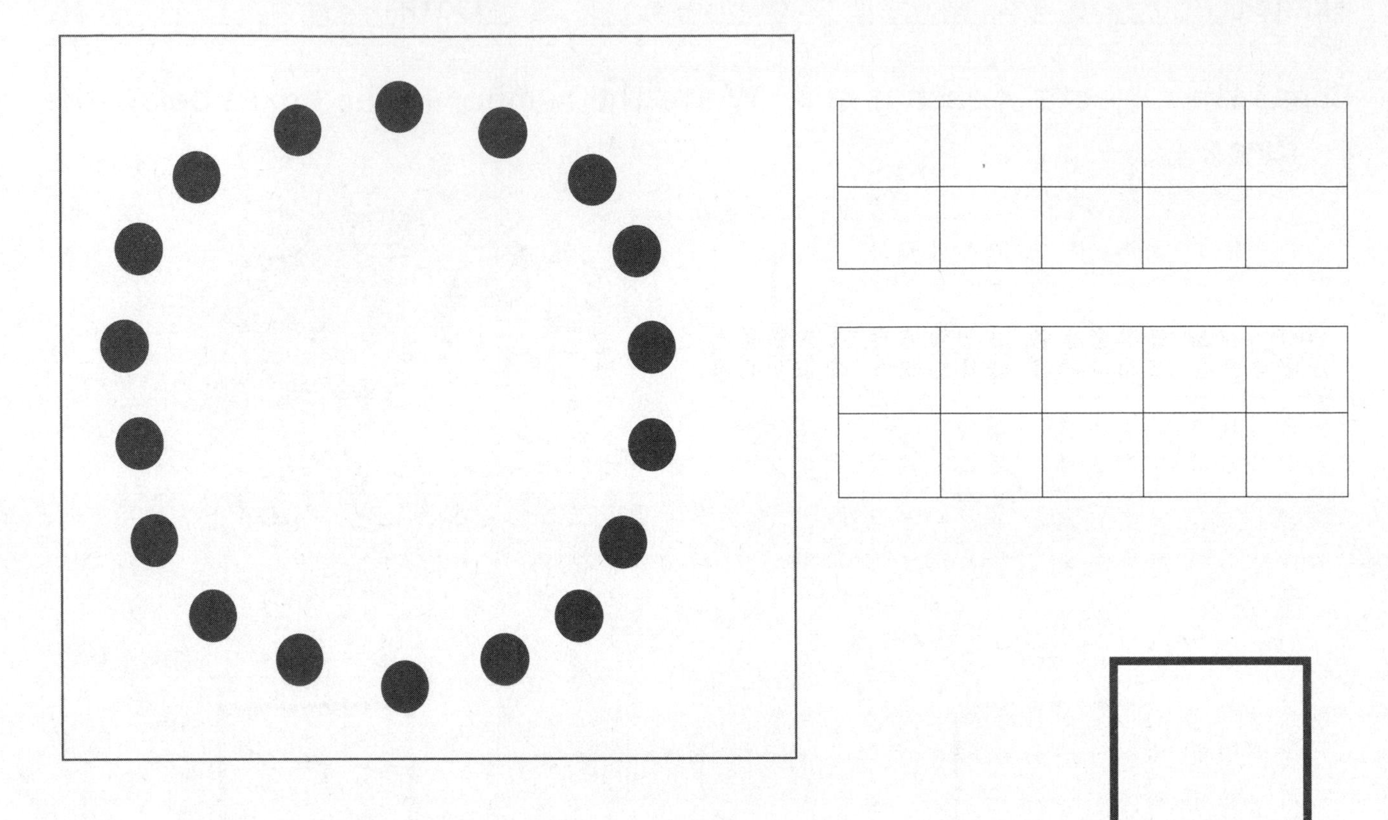

Write a teen number in the box below. Draw a picture to match your number.

Name ____________________ Date ______________

Count up by tens, and write the numbers.

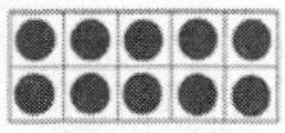	10
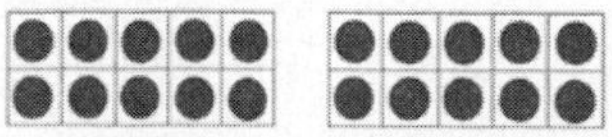	20
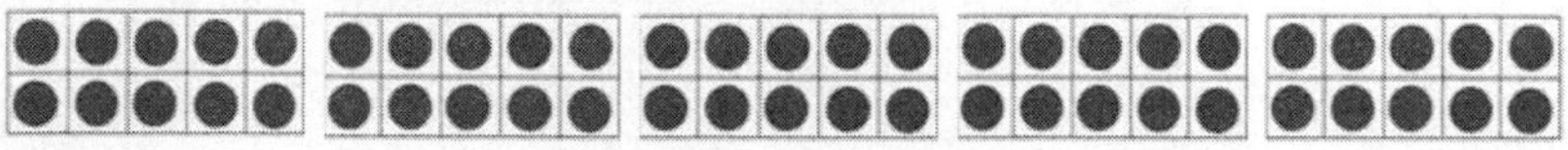	50

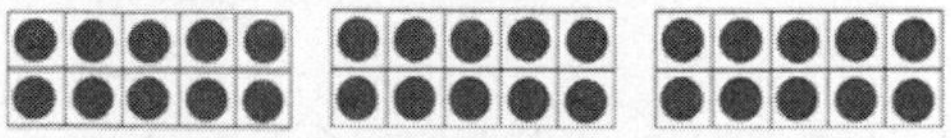

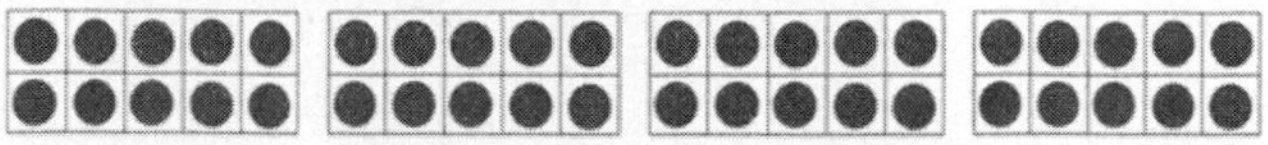

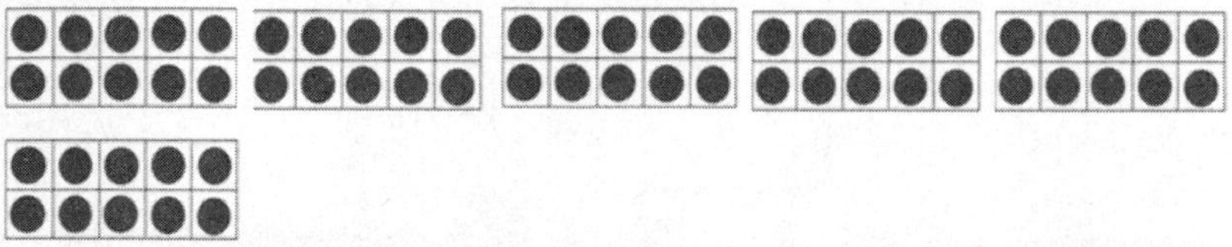

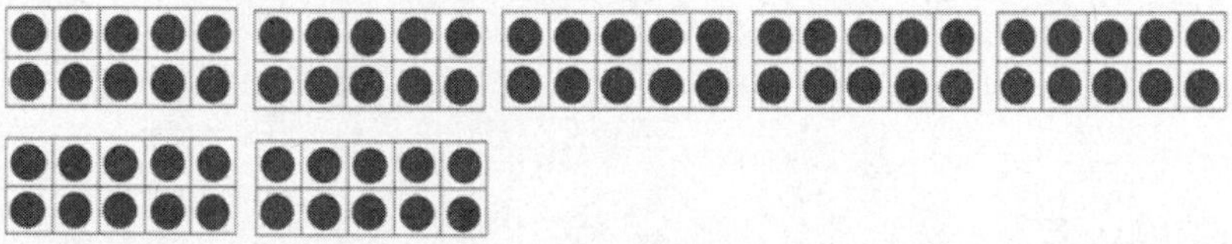

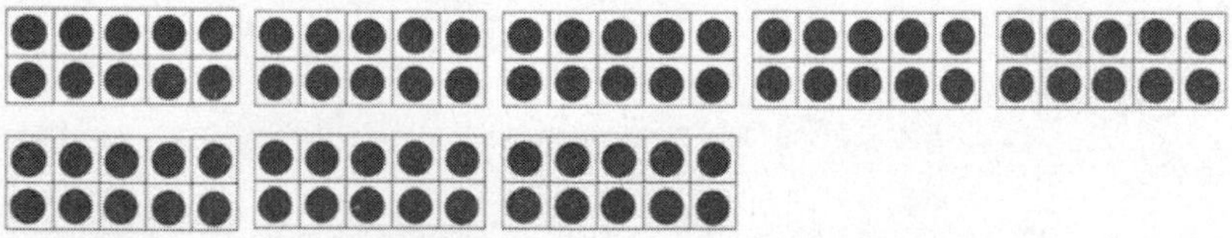

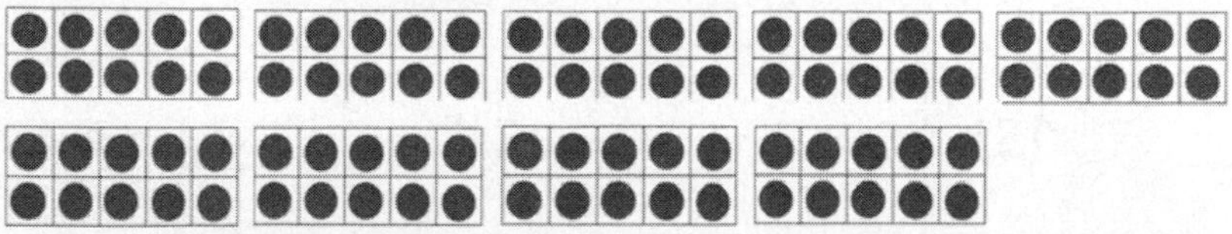

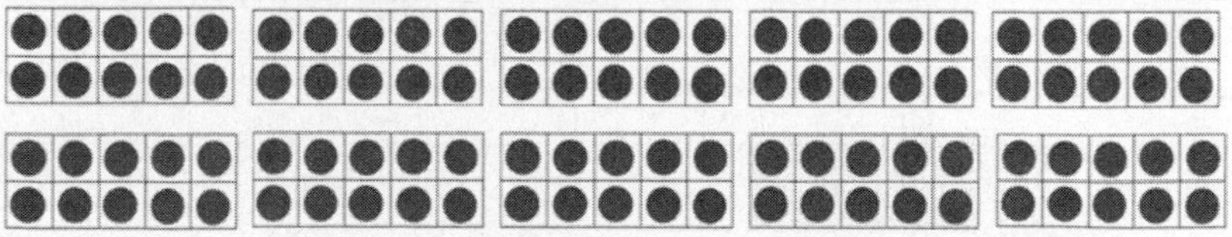

Help the puppy down the stairs! Count down by tens. Write the numbers.

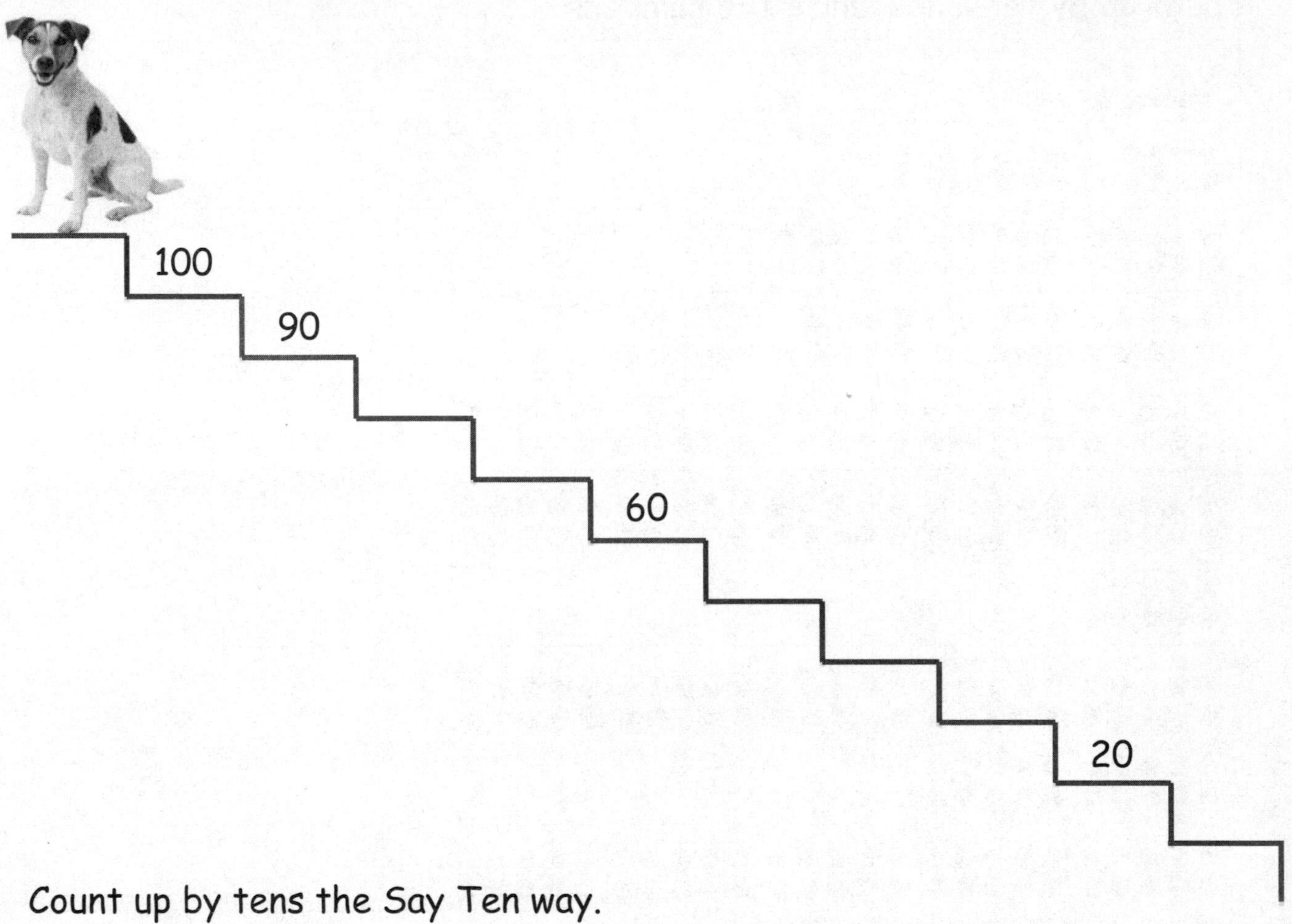

Count up by tens the Say Ten way.

ten	____ tens	3 tens	____ tens
____ tens	____ tens	____ ____	____ ____

Name ________________________ Date ____________

Count down by 10, and write the number on top of each stair.

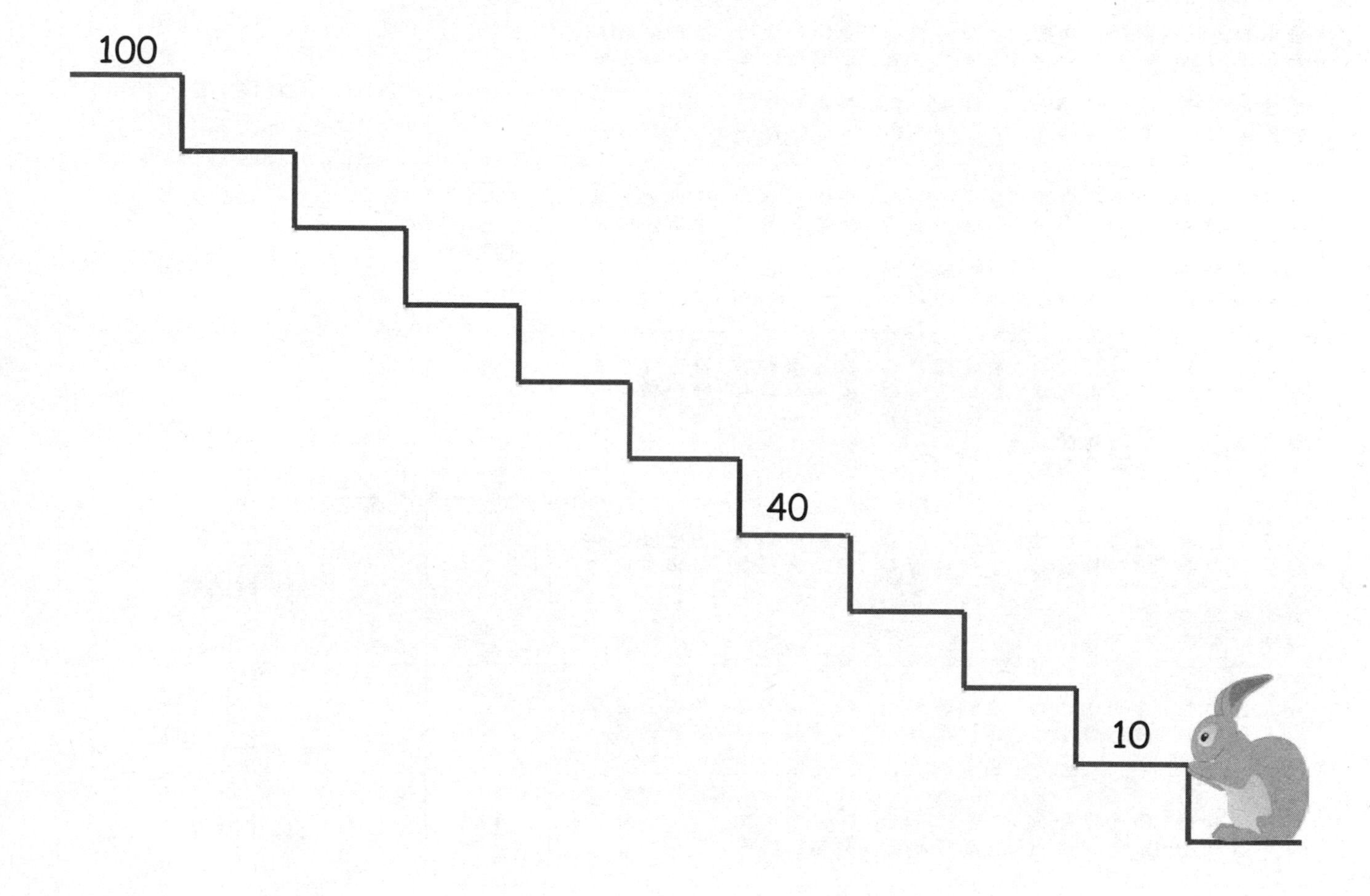

Count down the Say Ten way. Write the missing numbers.

	100	
		9 tens
	80	_____ tens
	70	_____ tens
		6 tens
		_____ tens
	40	4 tens
		_____ tens
		_____ tens
		_____ ten

Name ______________________________ Date _____________

Count up or down by 1s. Help the animals and the girl get what they want!

20 | ___ | 22 | ___ | 24 | ___ | 26 | ___ | ___ | ___

40 | ___ | ___ | ___ | 44 | ___ | 46 | ___ | 48

92 | ___ | ___ | ___ | ___ | ___ | 98 | 99

Count up.

63	64			

Stop!

Count down.

66			

Name ______________________ Date ____________

Help the rabbit get his carrot. Count by 1s.

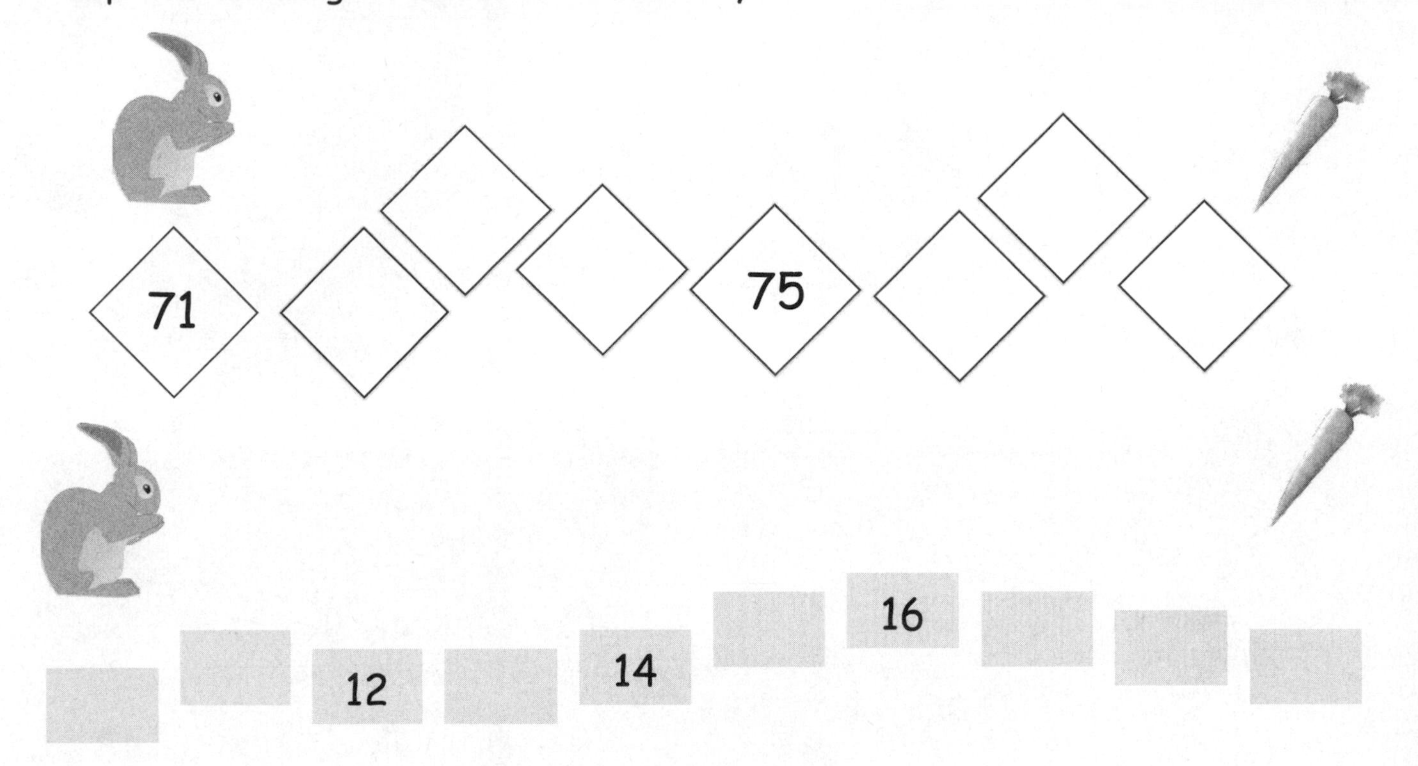

Count up by 1s, then down by 1s.

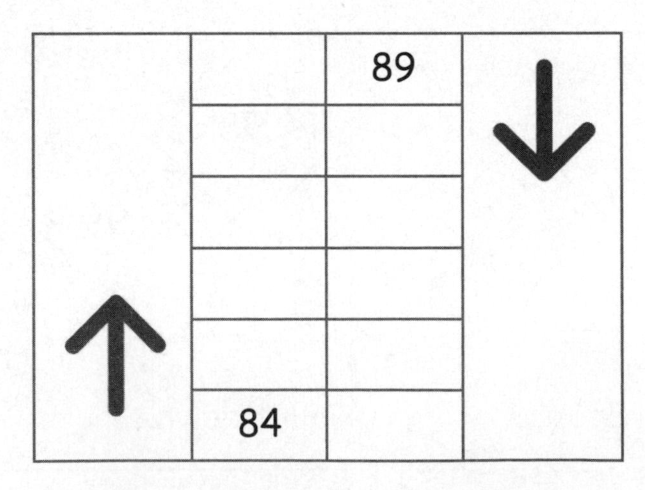

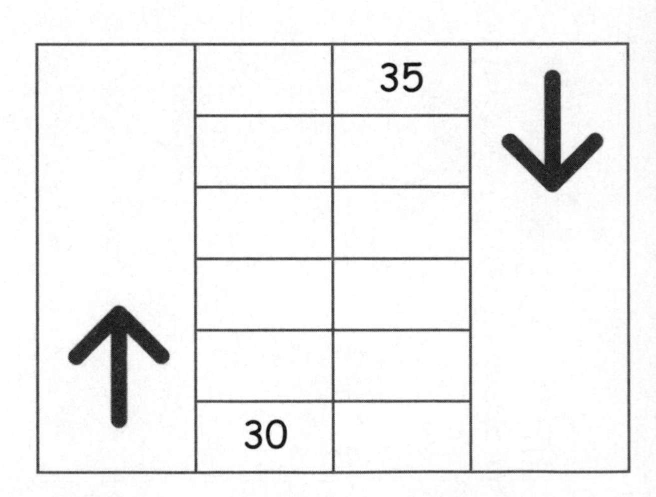

Help the boy mail his letter. Count up by 1s. When you get to the top, count down by 1s.

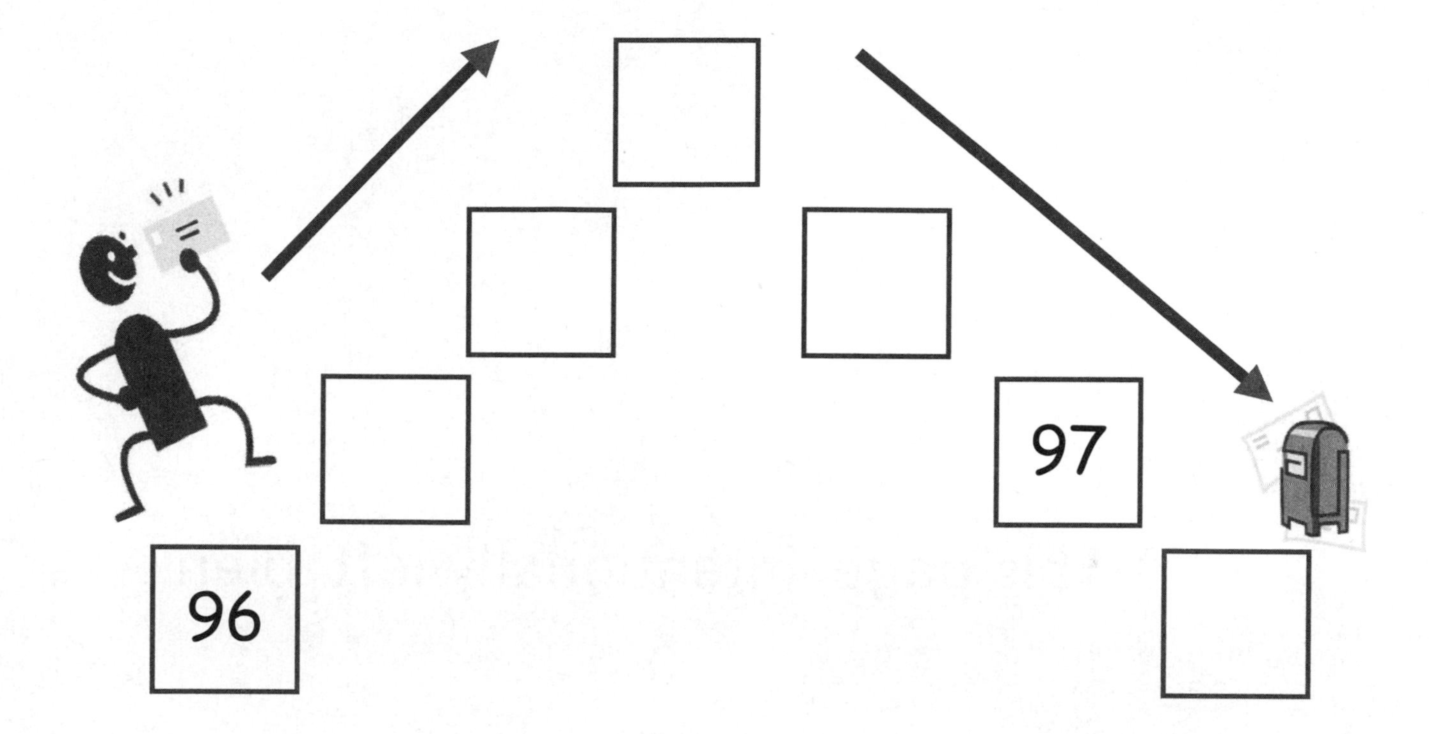

This page intentionally left blank

Name ______________________________ Date ______________

Touch and count the dots from left to right starting at the arrow. Count to the puppy, and then keep counting to his bones and twin brother!

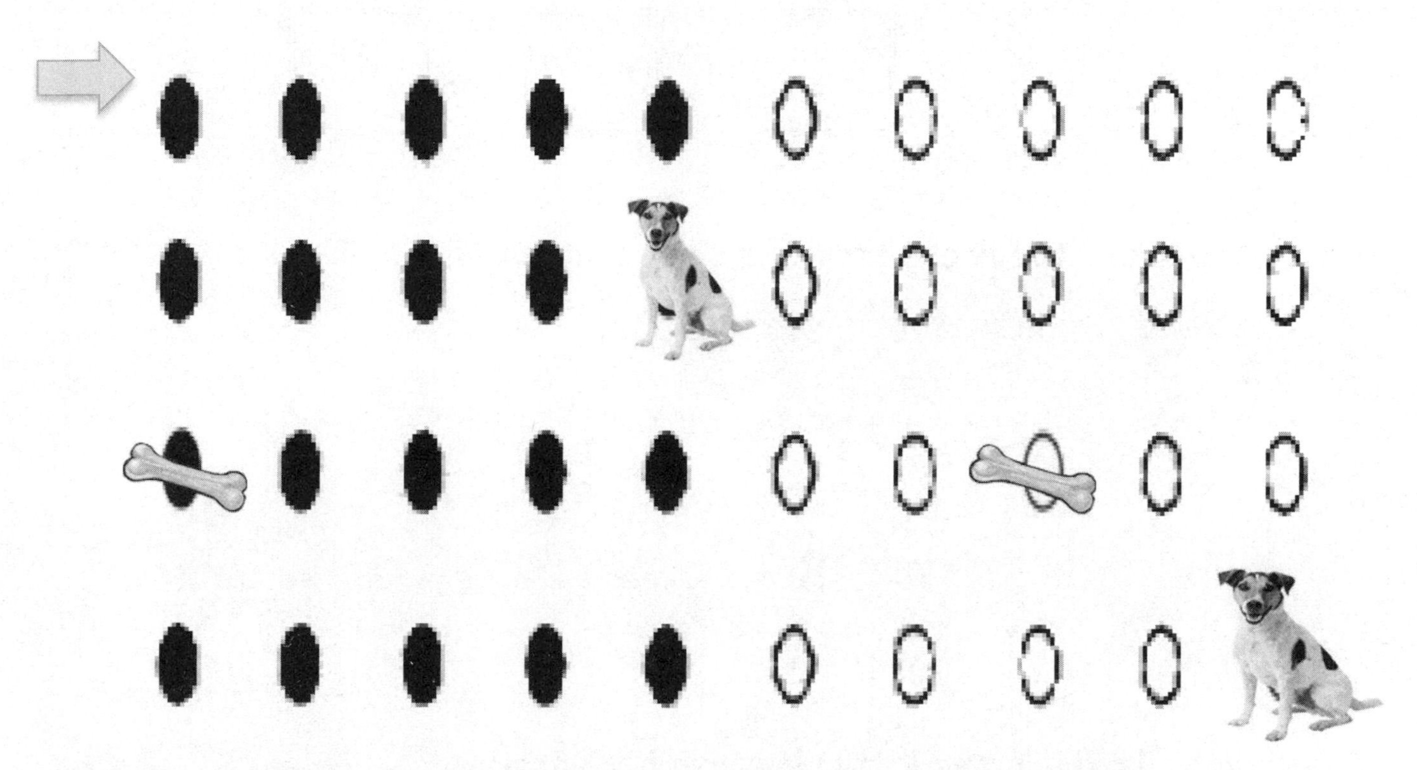

Count again and color the last dot of each row green. When you have finished, go back and see if you can remember your green numbers!

What number did you say when you touched the first puppy?

- The first bone?
- The second bone?
- His twin brother?

Count each number by 1s. Write the number below when there is a box.

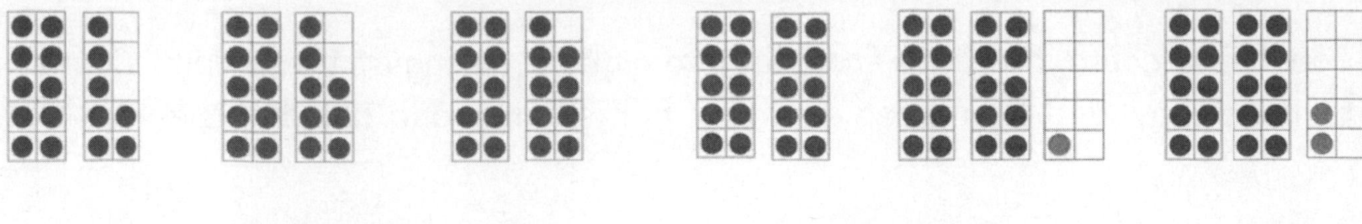

17				21	

Touch and count the rocks from the cow to the grass!

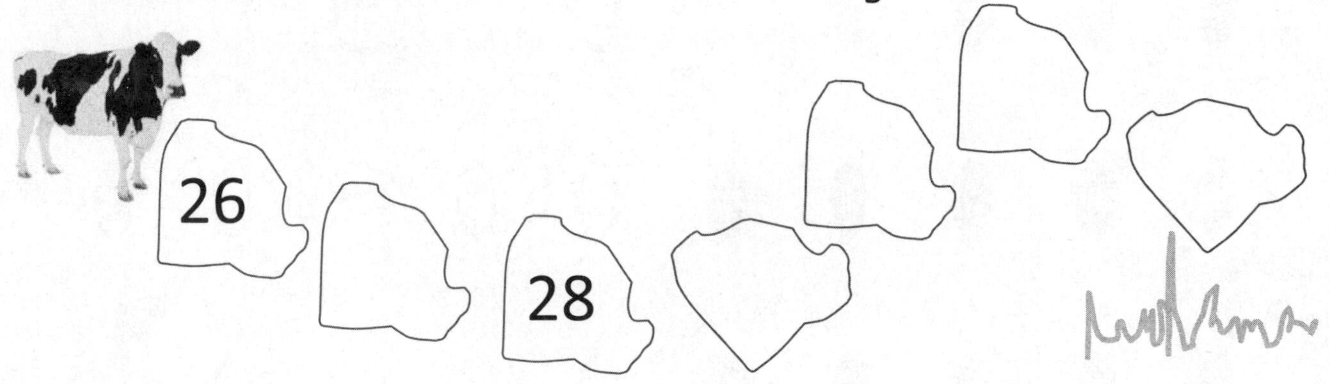

Count up by 1s. Help the kitty play with her yarn!

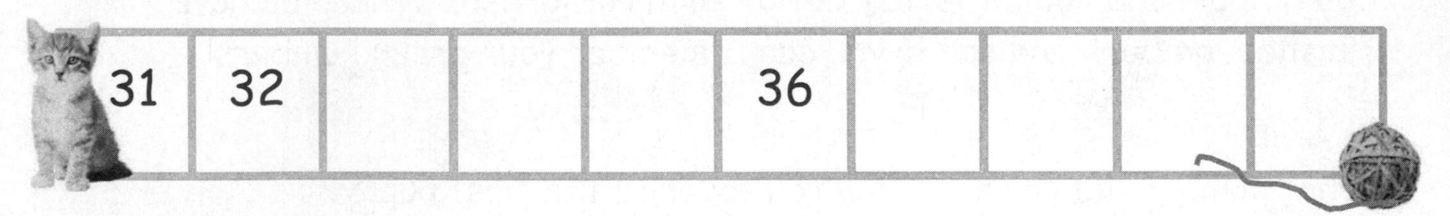

Count down by 1s.

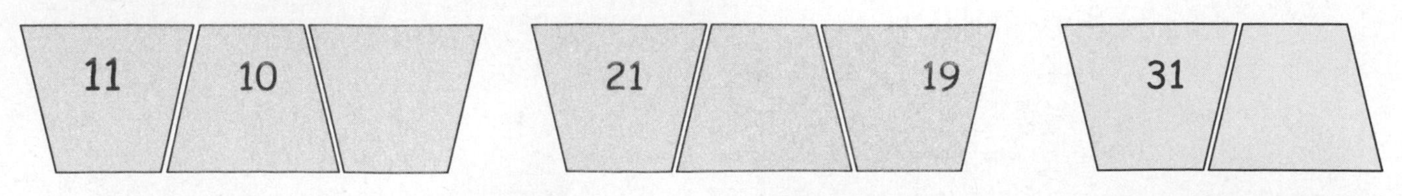

Name ______________________ Date ____________

Draw more to show the number.

Example: 5

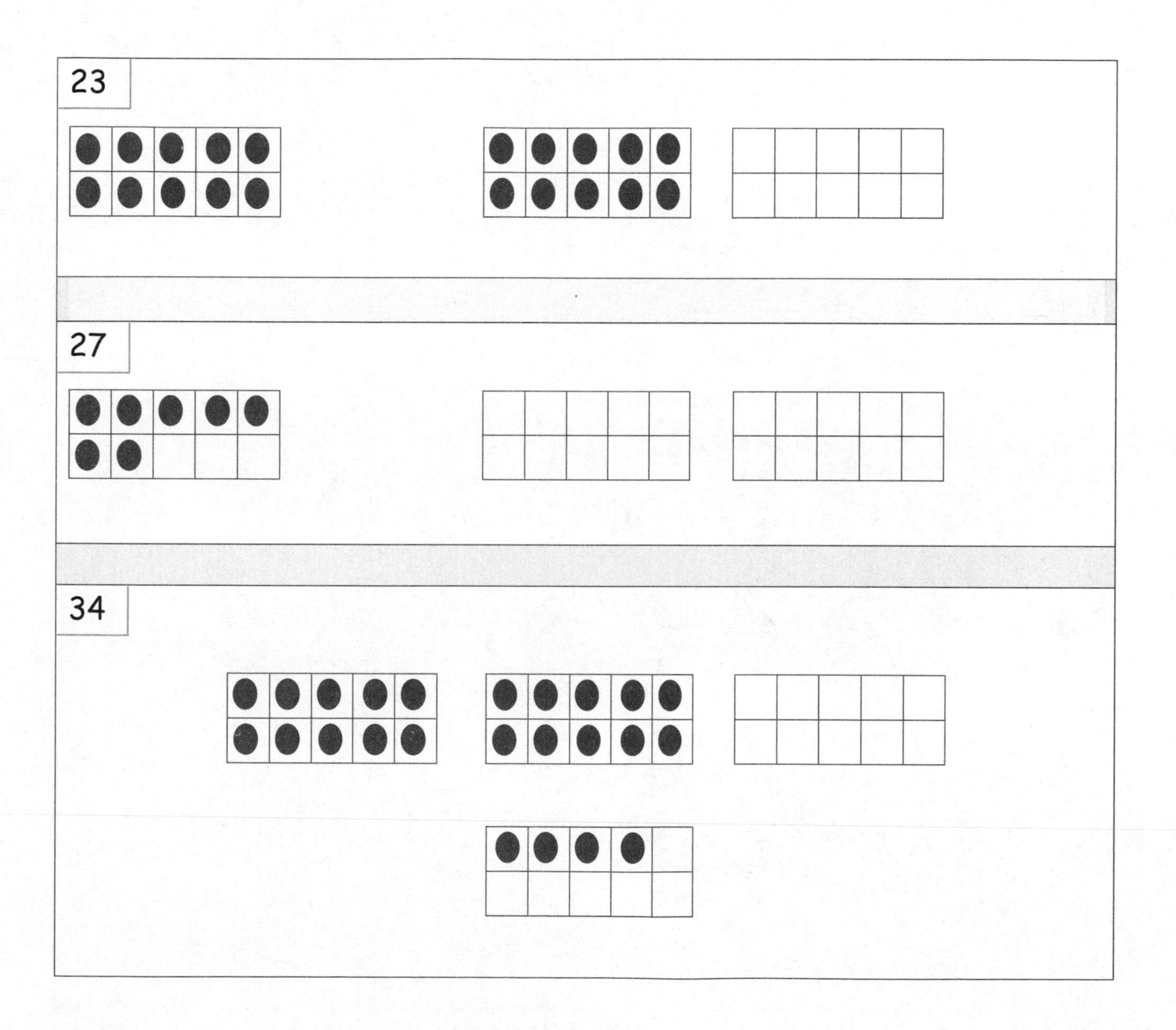

38

40

Name ______________________________ Date ______________

Rekenrek

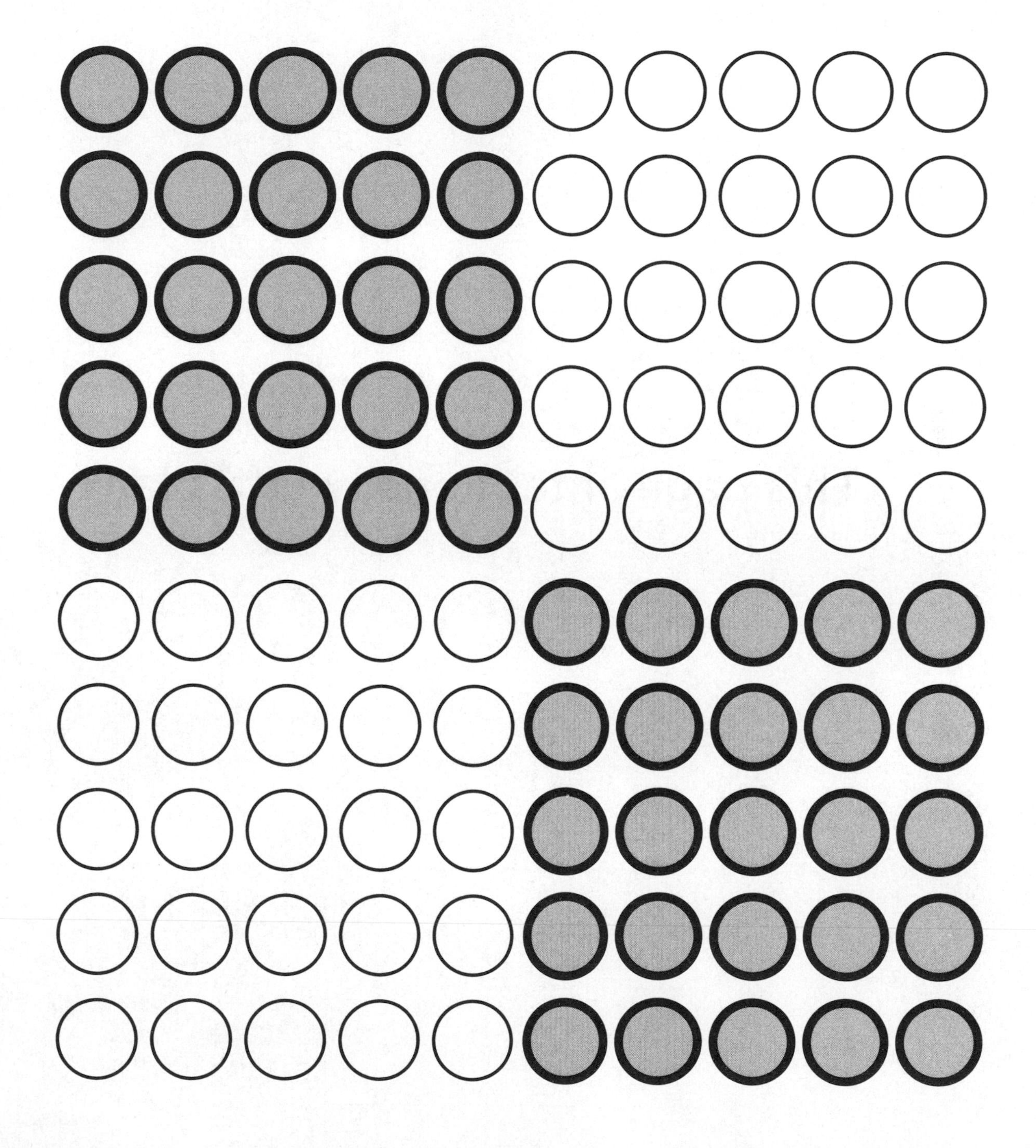

This page intentionally left blank

Directions for Rekenrek Homework

Use your Rekenrek (attached), hiding paper (an extra paper to hide some of the dots), and crayons to complete each step listed below. Read and complete the problems with the help of an adult.

Hide to show just 40 on your Rekenrek dot paper. Touch and count the circles until you say 28. Color 28 green.

- Touch and count each circle from 28 to 34.
- Color 34 (the 34th circle) with a red crayon.

Hide to show just 60 on your Rekenrek dot paper. Touch and count the circles until you say 45. Color 45 yellow.

- Touch and count each circle from 45 to 52.
- Color 52 with a blue crayon.

Hide to show just 90 on your Rekenrek dot paper. Touch and count the circles until you say 83. Color 83 purple.

- Touch and count down from 83 to 77.
- Color 77 with a red crayon.

Show 100.

- Touch and count, starting at 1.
- Say the last number in each row loudly. Color the circle black.

Name ______________________________ Date ______________

Rekenrek

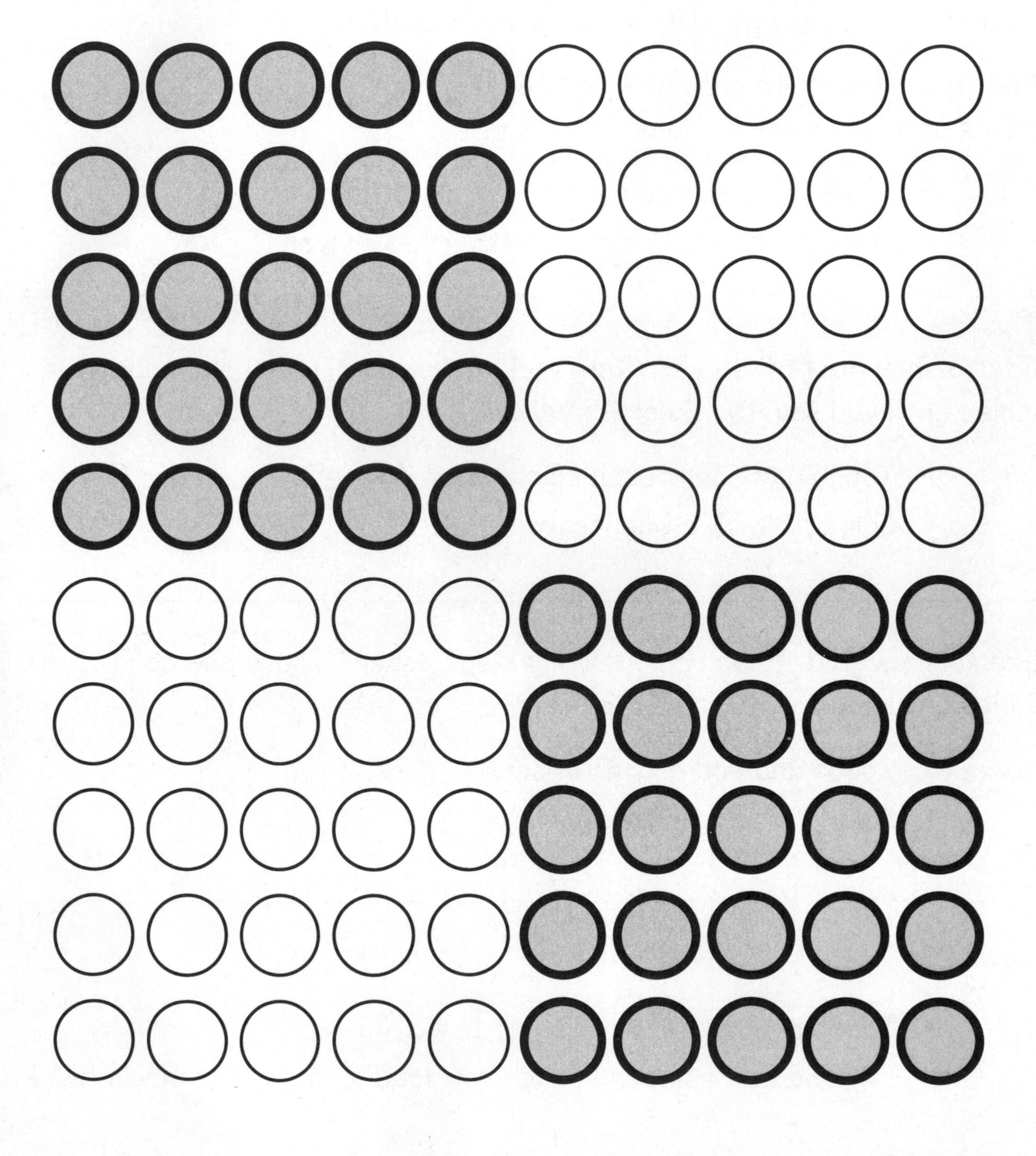

Name ______________________________ Date ______________

Find the Hidden Teen Number

Show each number on your Rekenrek with your partner. Write how many. Circle the teen number inside the big number. Draw a line from the big number to the teen number that hides inside it.

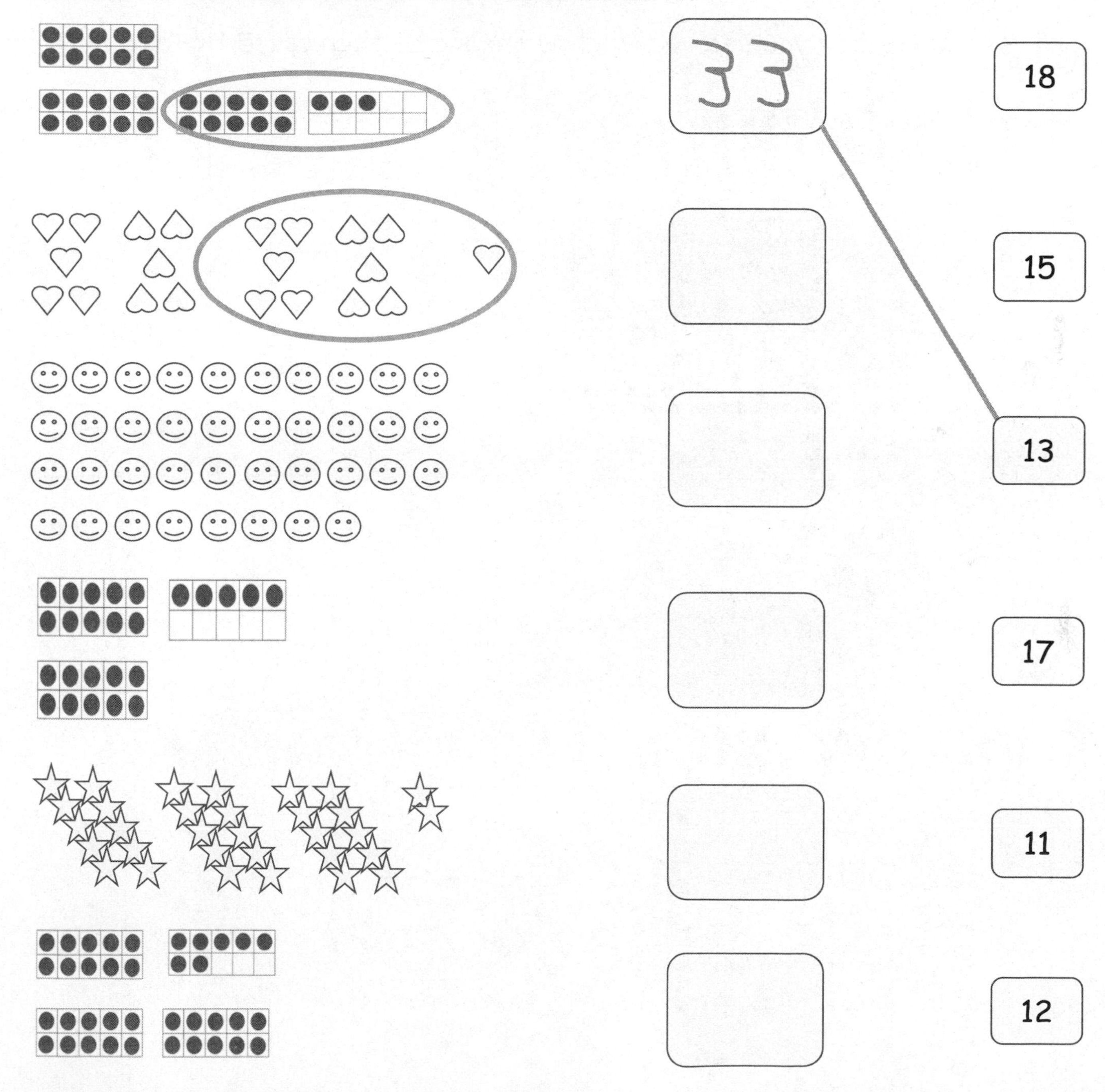

Name ______________________________ Date ______________

Write the number you see. Now, draw one more, then write the new number.

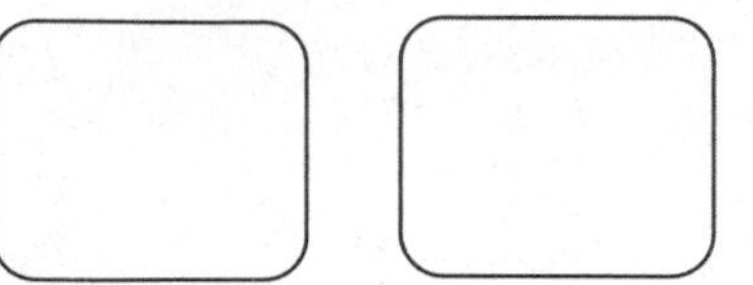
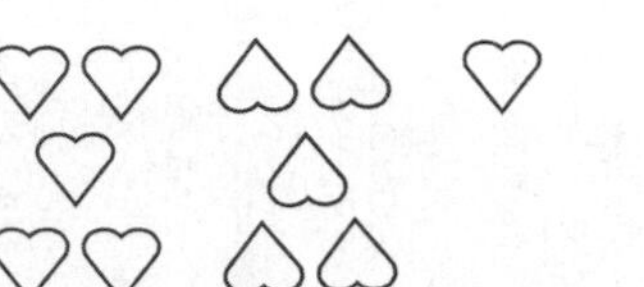
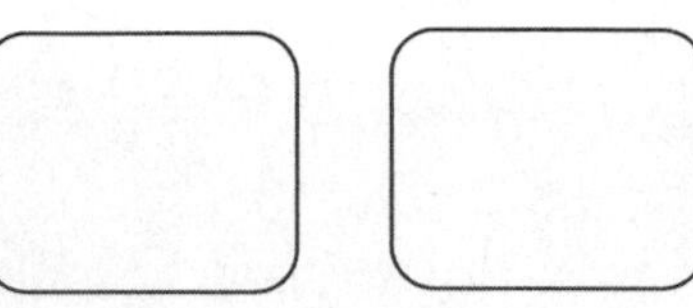
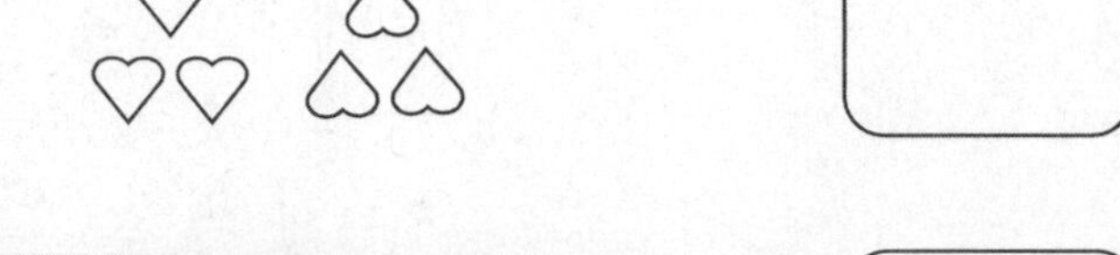
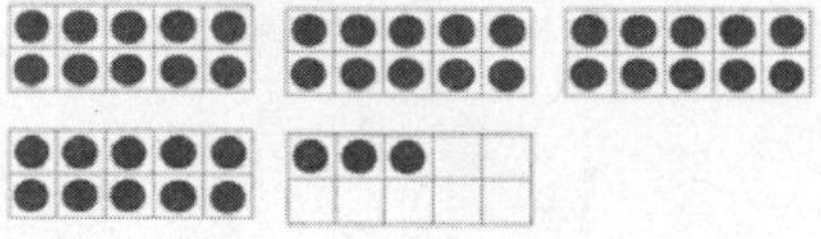
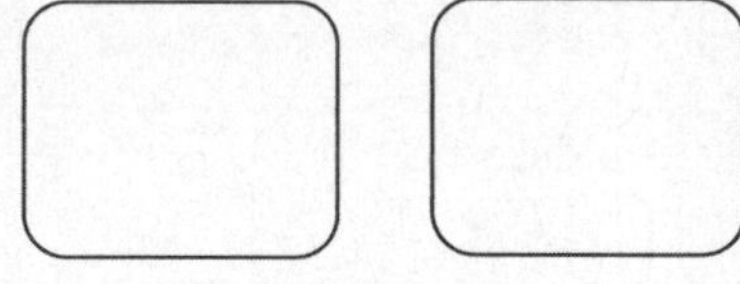

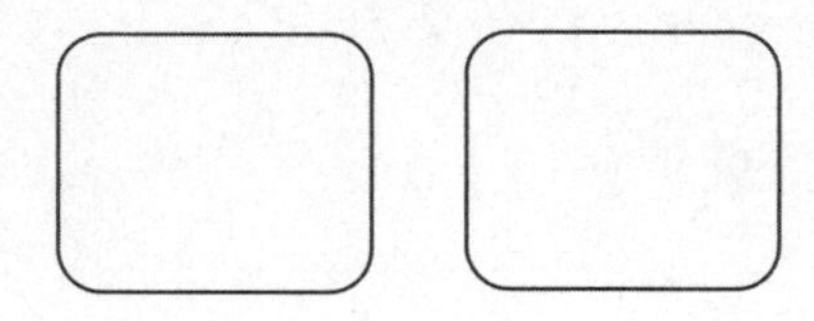
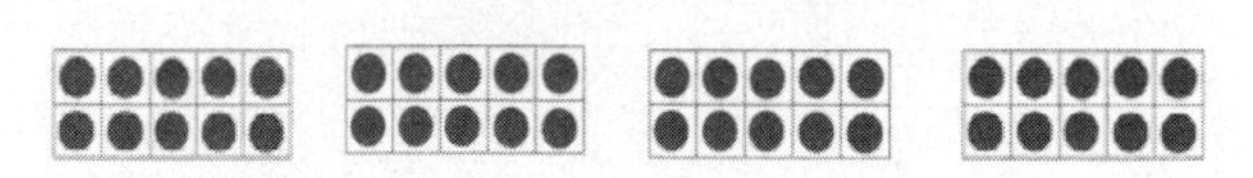
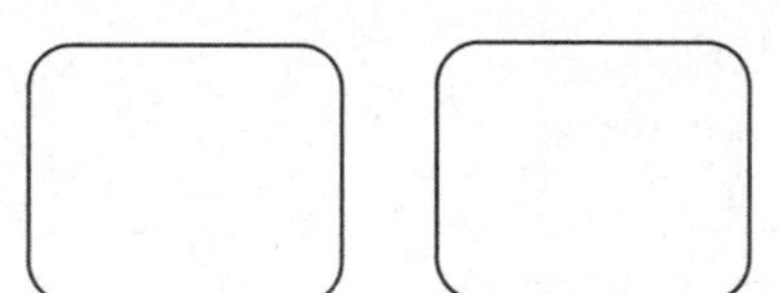

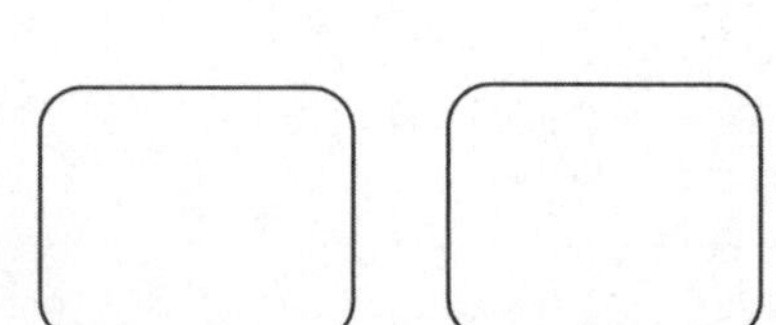

EUREKA MATH™

Name ______________________________ Date ______________

Fill in each number bond, and write a number sentence to match.

Example:

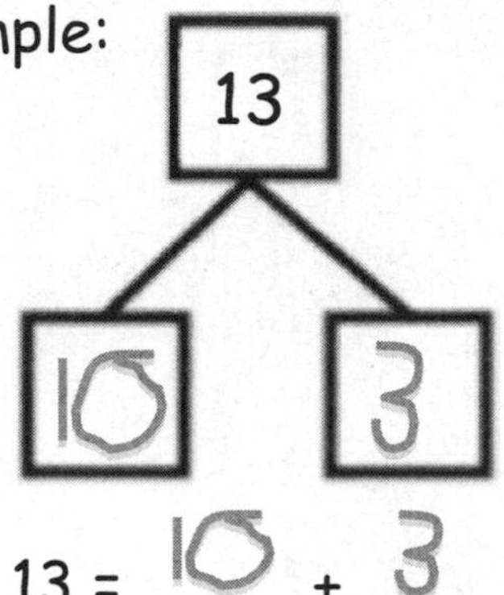

13 = 10 + 3

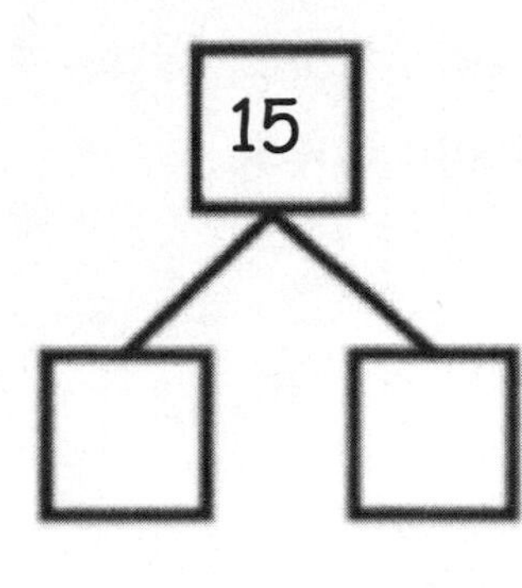

15 = _____ + _____

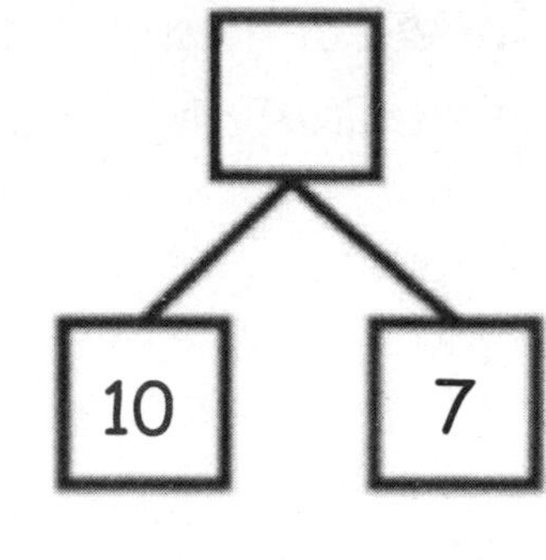

17 = _____ + _____

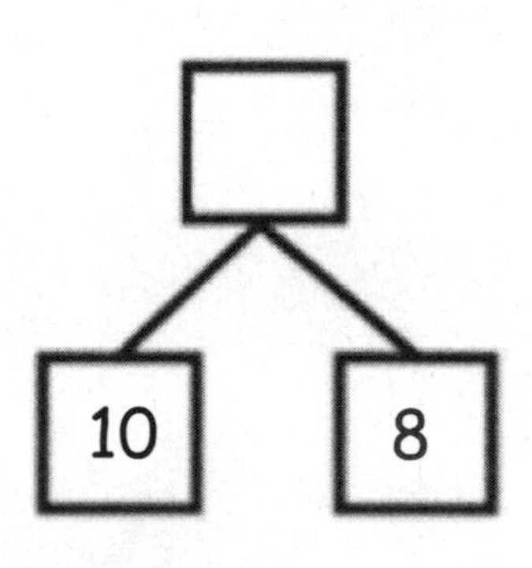

10 + 8 = _____

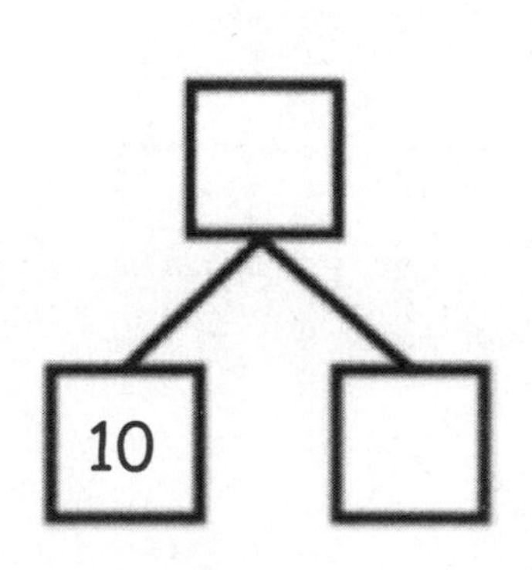

10 + 6 = _____

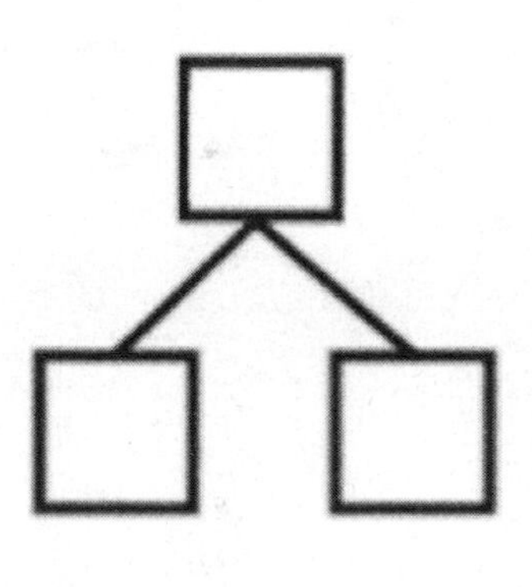

_____ = 10 + 4

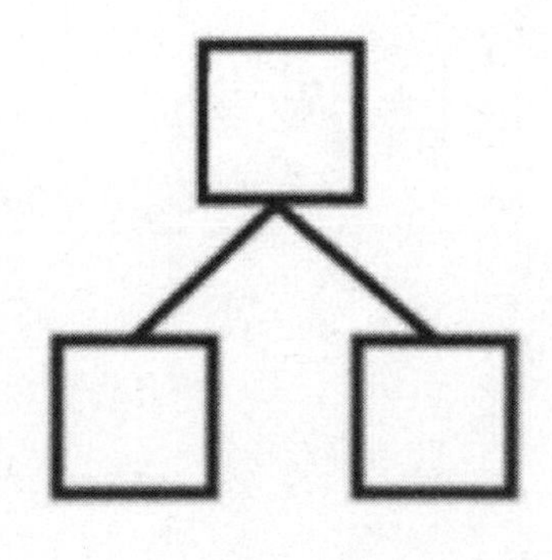

12 = _____ + _____

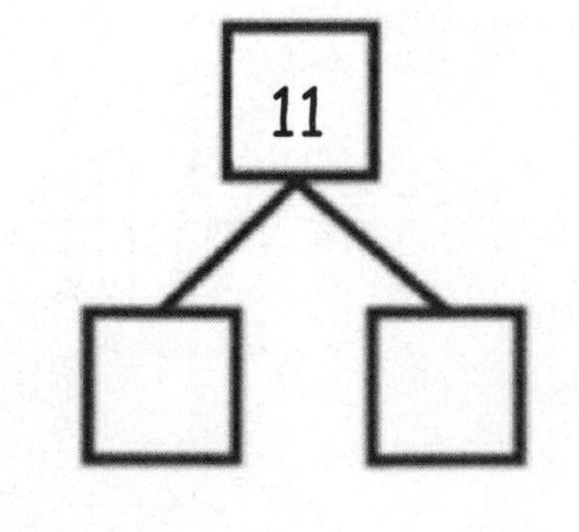

_____ = _____ + _____

Early finishers: Make up your own teen number bonds and number sentences on the back!

Name ______________________________ Date ______________

Draw stars to show the number as a number bond of 10 ones and some ones. Show each example as two addition sentences of 10 ones and some ones.

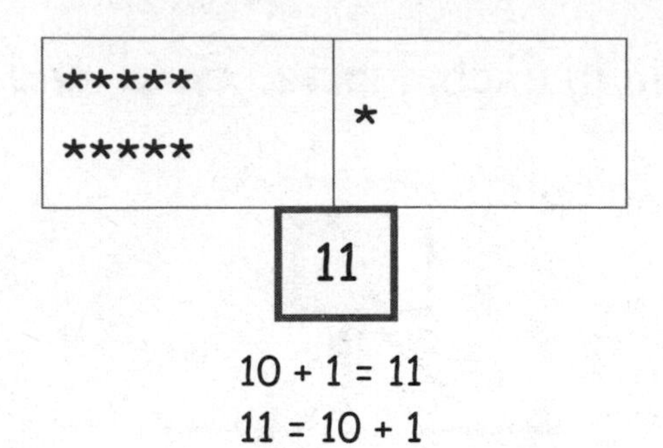

* * * * * * * * * *	

15

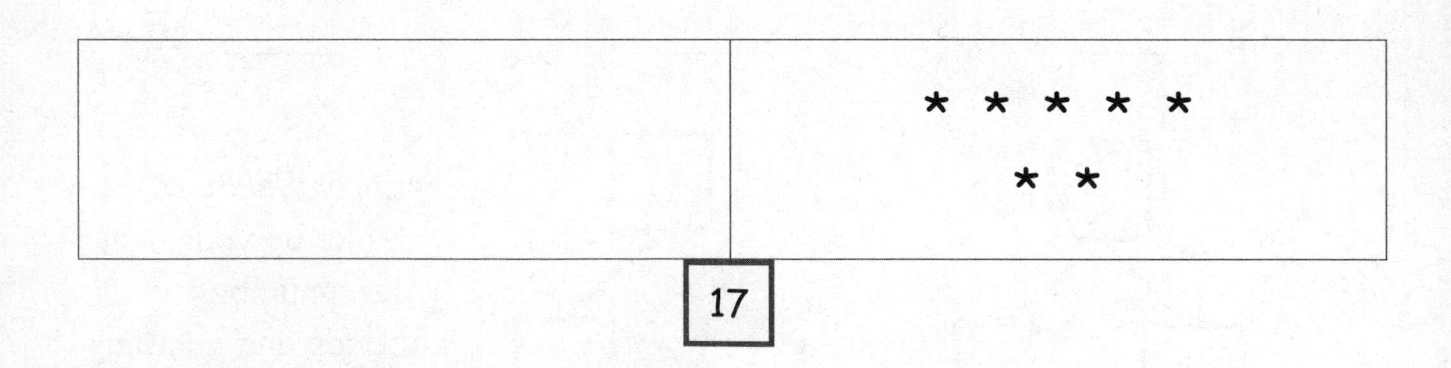

EUREKA MATH™

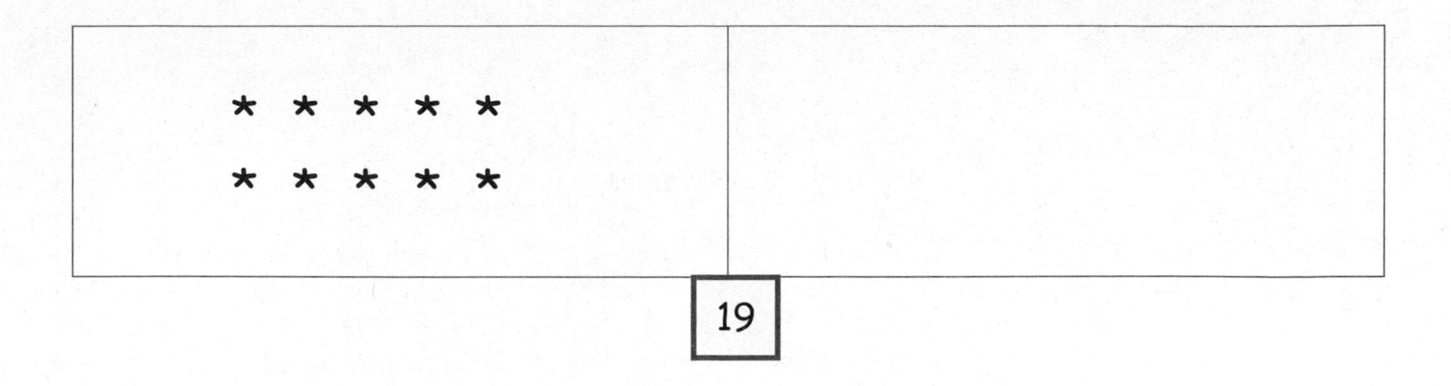

14

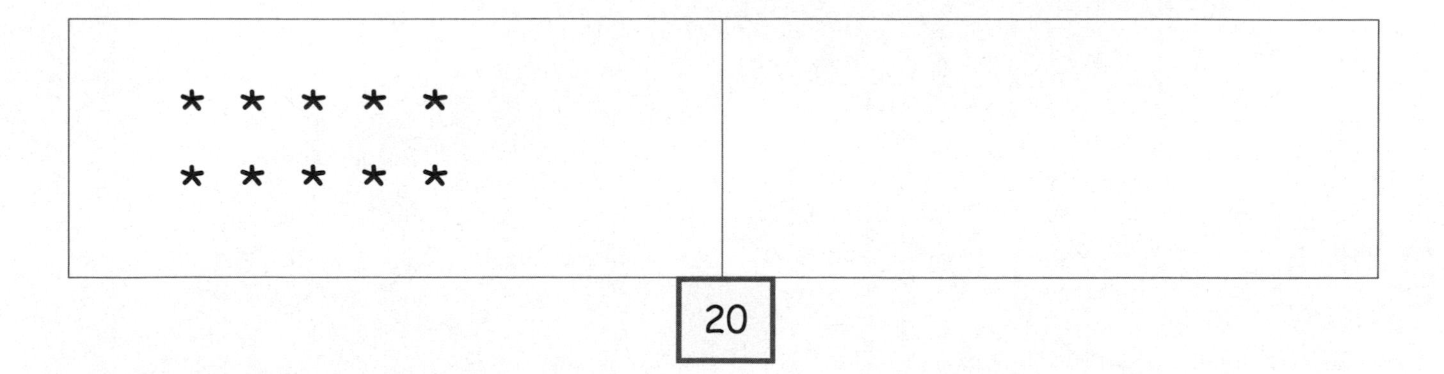

This page intentionally left blank

Name ______________________________ Date ______________

Model each number with cubes on your number bond mat. Then, complete the number sentences and number bonds.

Example:

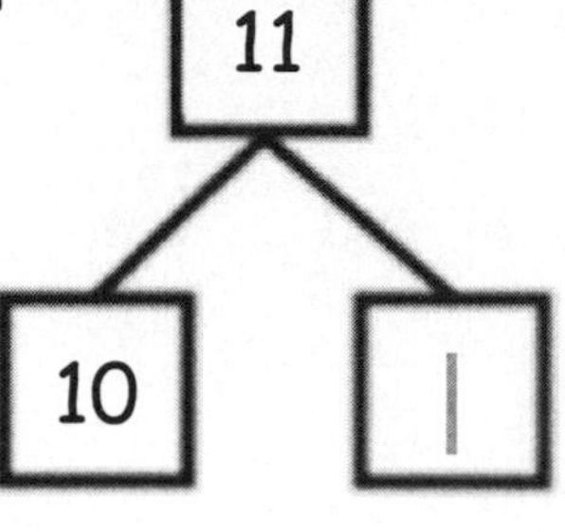

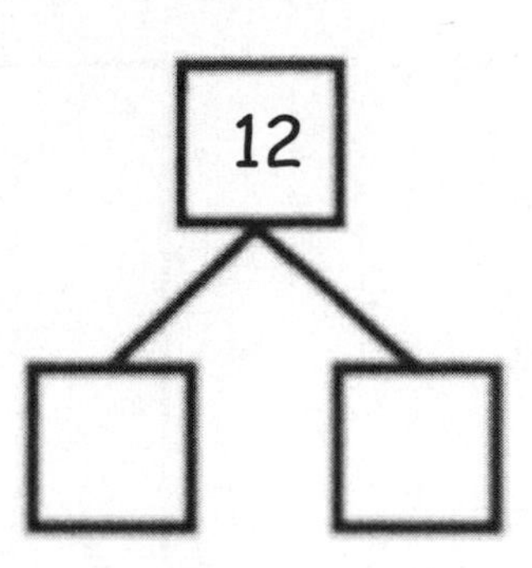

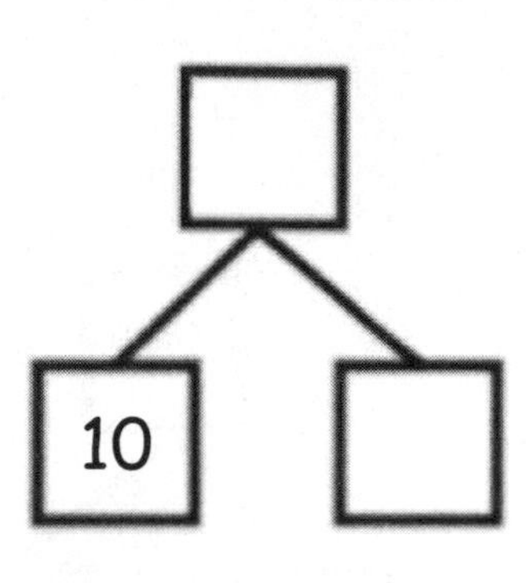

11 = 10 + __|__ 12 = 10 + _____ 13 = 10 + _____

10 + __|__ = 11 10 + _____ = 12 10 + _____ = 13

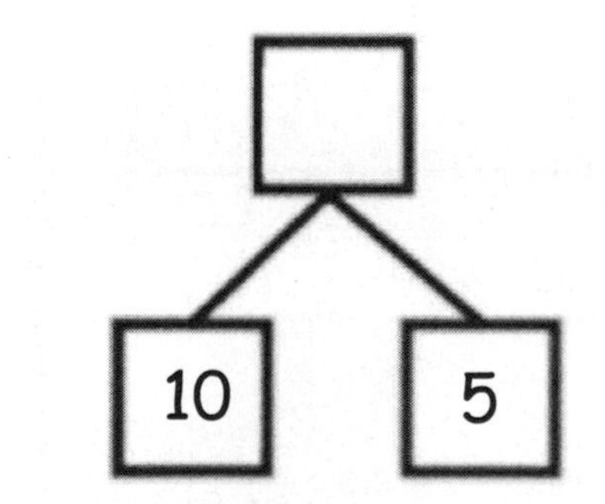

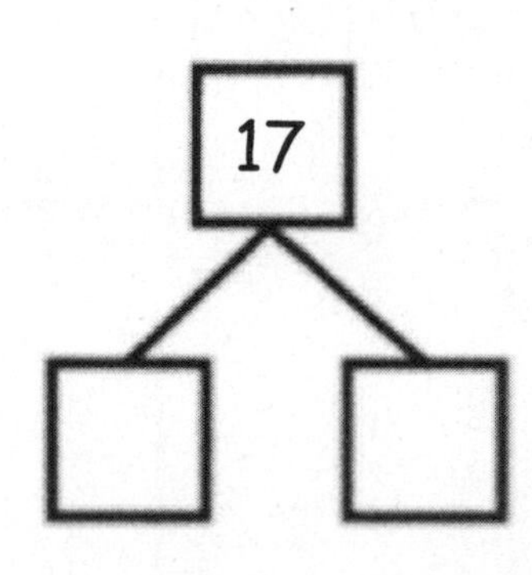

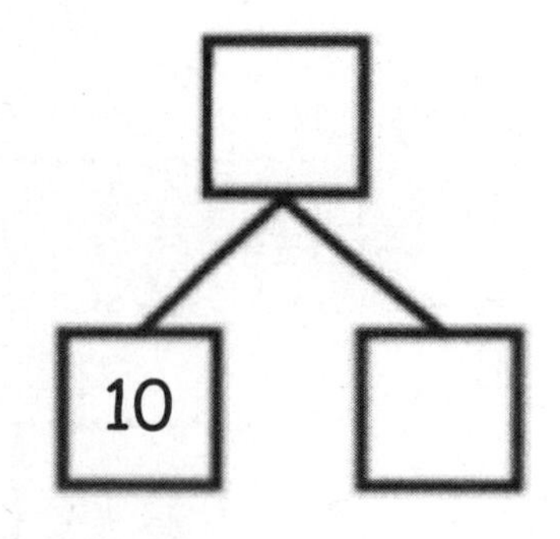

_____ + 5 = 15 _____ + 7 = 17 _____ + 8 = 18

15 = _____ + 5 17 = _____ + 7 18 = 10 + _____

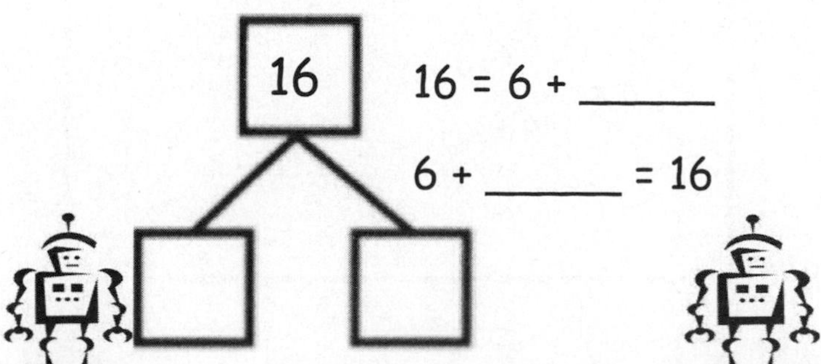

16 = 6 + _____

6 + _____ = 16

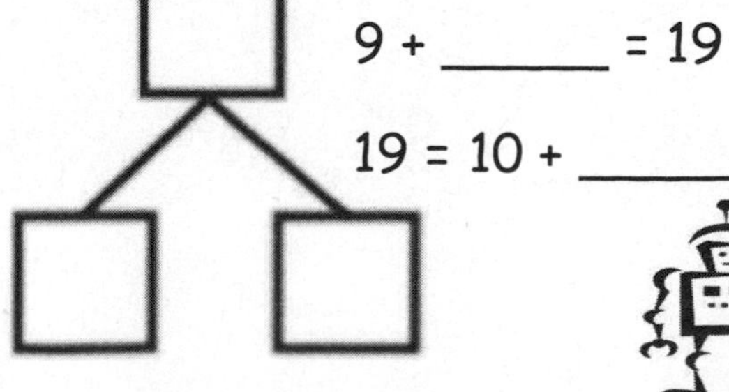

9 + _____ = 19

19 = 10 + _____

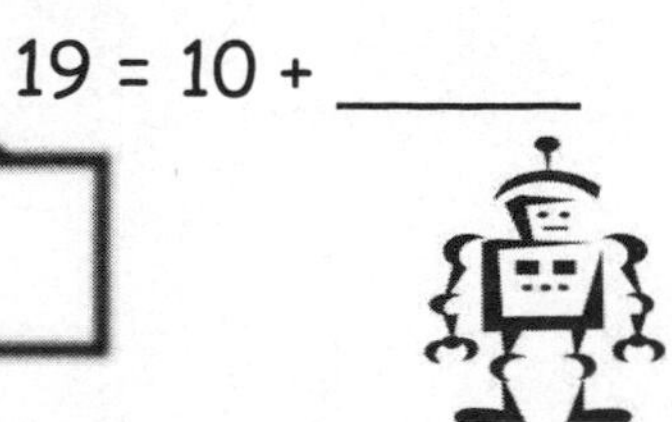

Name ______________________________ Date ______________

Complete the number bonds and number sentences. Draw the squares of the missing part.

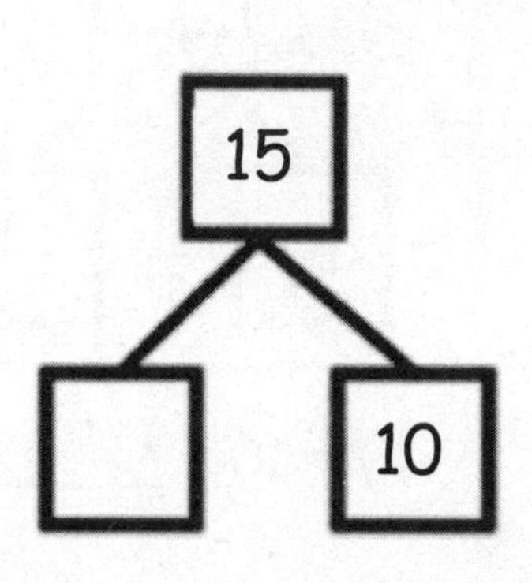

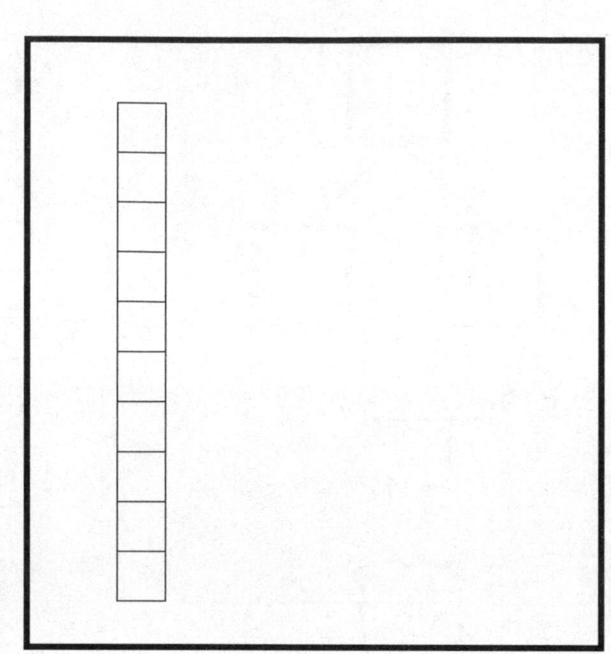

15 = _______ + 10

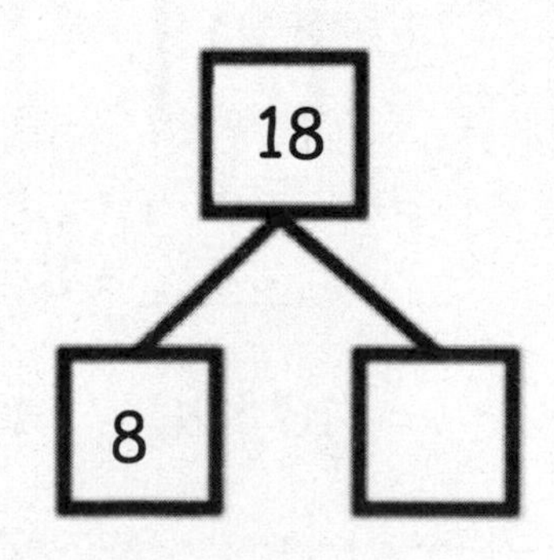

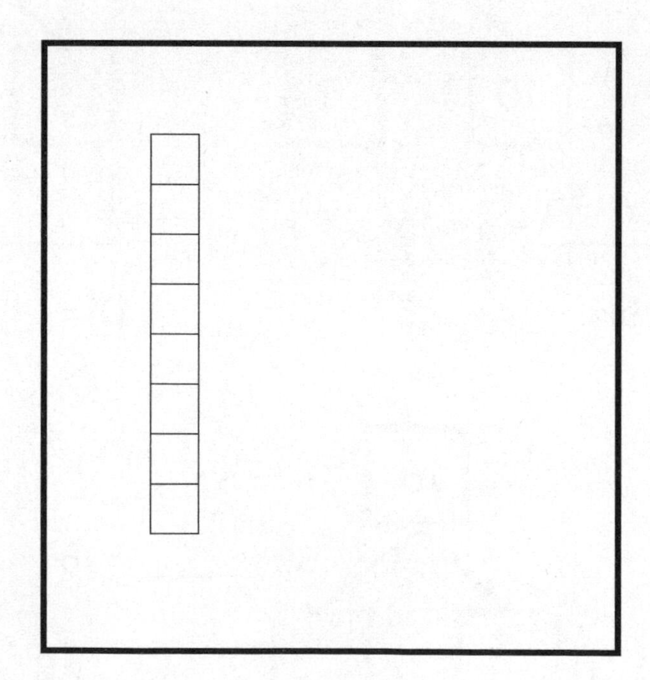

_______ + 8 = 18

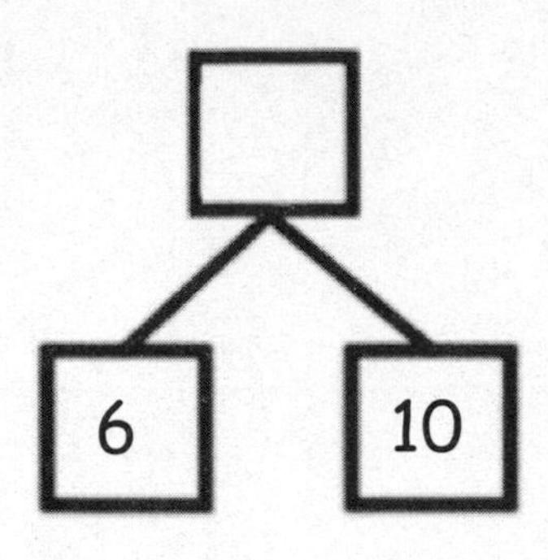

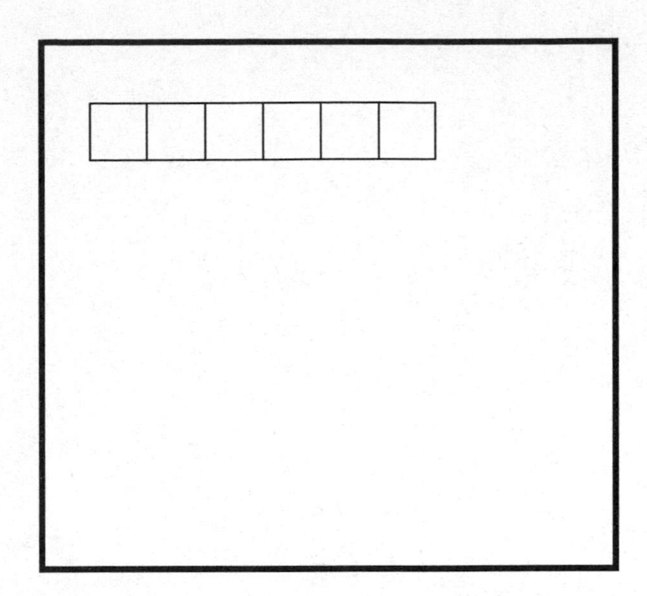

6 + ______ = 16

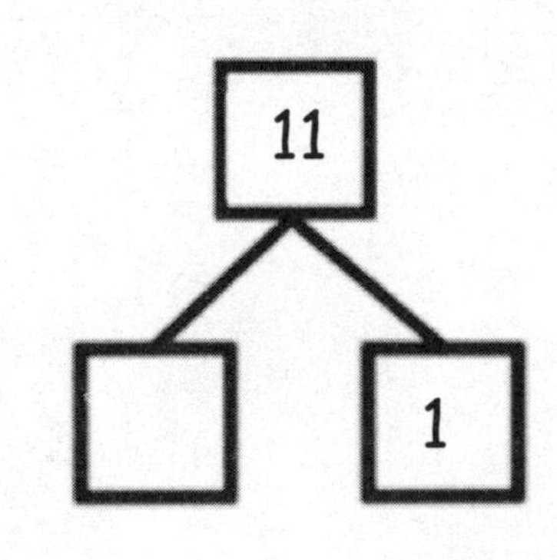

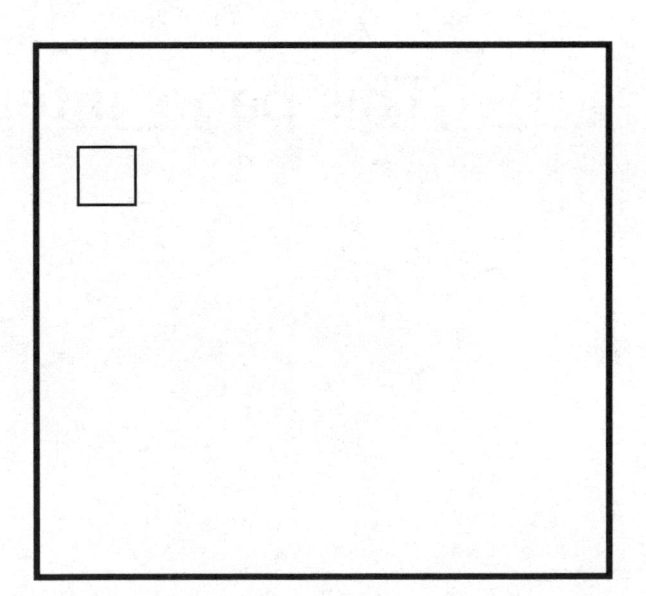

1 + ______ = 11

EUREKA MATH
Lesson 21: Represent teen number decompositions as 10 ones and some ones, and find a hidden part.
73

©2015 Great Minds. eureka-math.org
GK-M5-SE-B4-1.3.1-01.2016

This page intentionally left blank

Name ______________________________ Date ______________

Circle 10 erasers. Circle 10 pencils. Match the extra ones to see which group has more. ✓ Check the group that has *more* things.

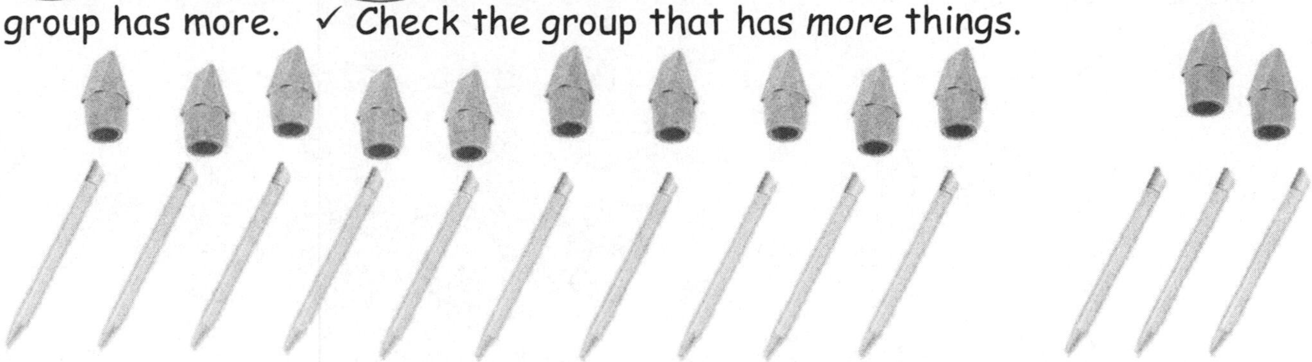

Circle 10 sandwiches. Circle 10 milk cartons. ✓ Check the group that has *less* things.

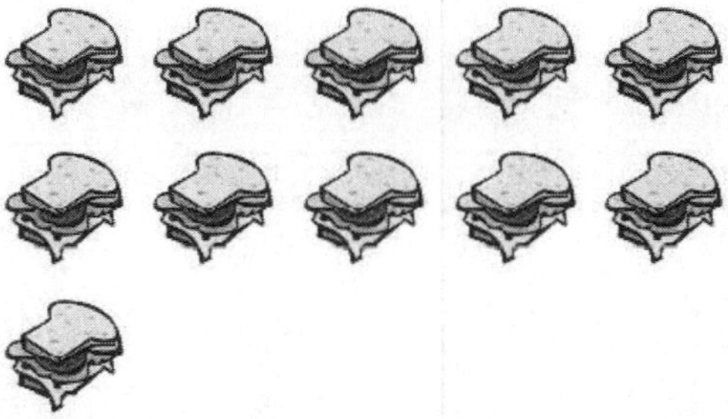

Circle 10 baseballs. Circle 10 gloves. Write how many are in each group. ✓ Check the group that has *more* things.

Circle 10 apples. Circle 10 oranges. Write how many are in each group.
✓ Check the group that has *less*.

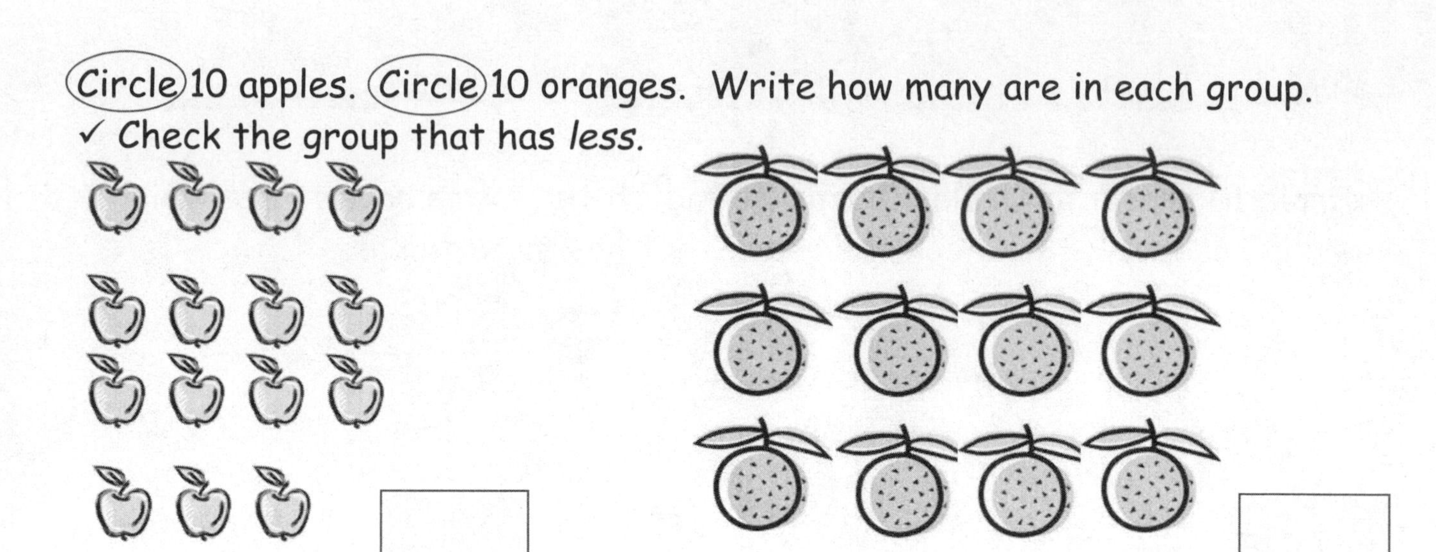

Circle 10 spoons. Circle 10 forks. Write how many are in each group.
Circle *more* or *less*.

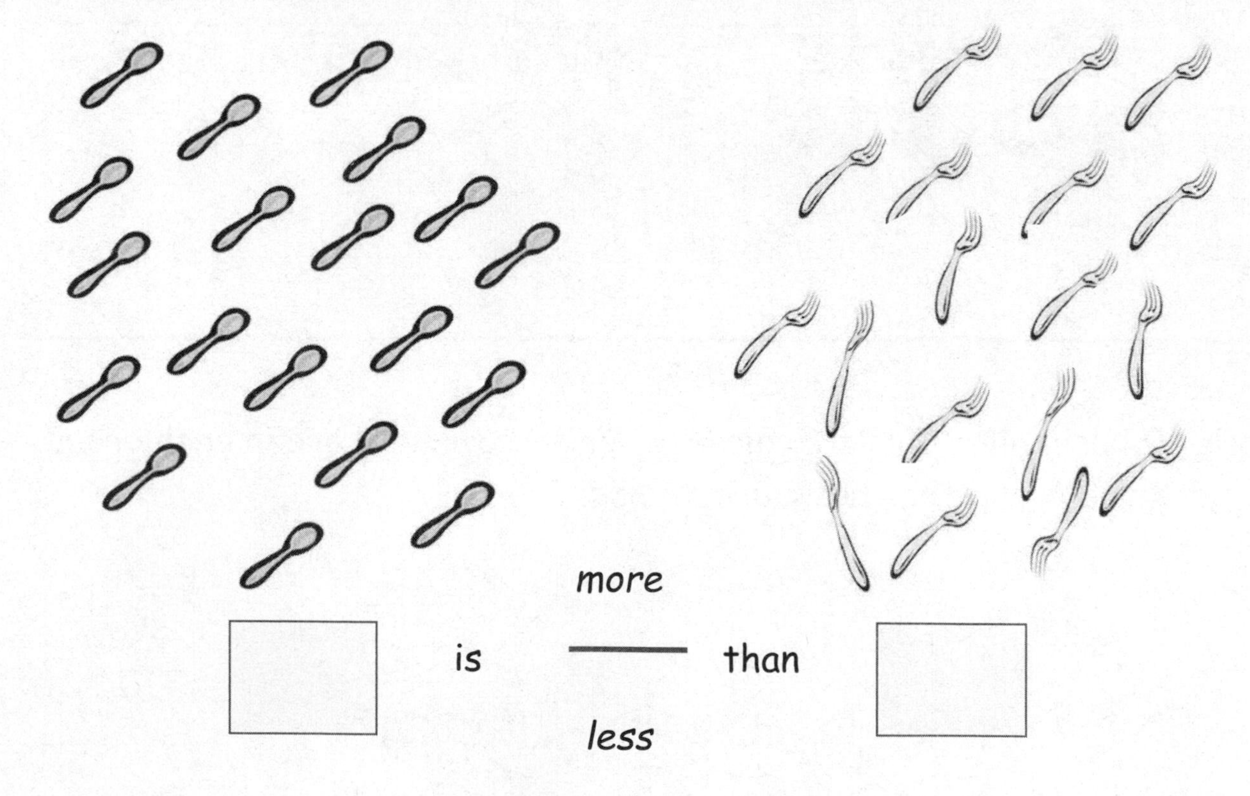

more

☐ is ____ than ☐

less

Name ______________________________ Date ____________

Fill in the number bond.
Check the group with *more*.

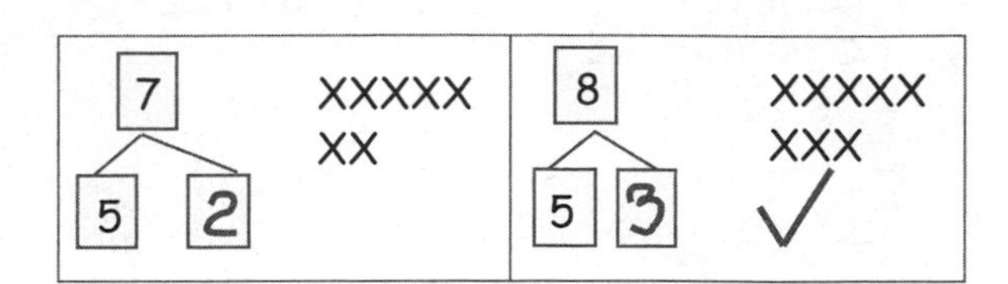

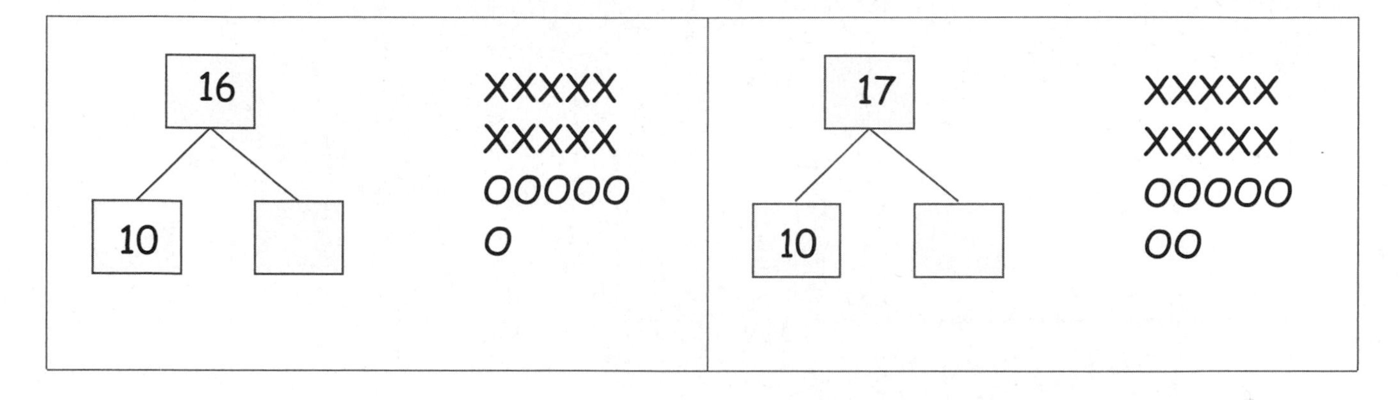

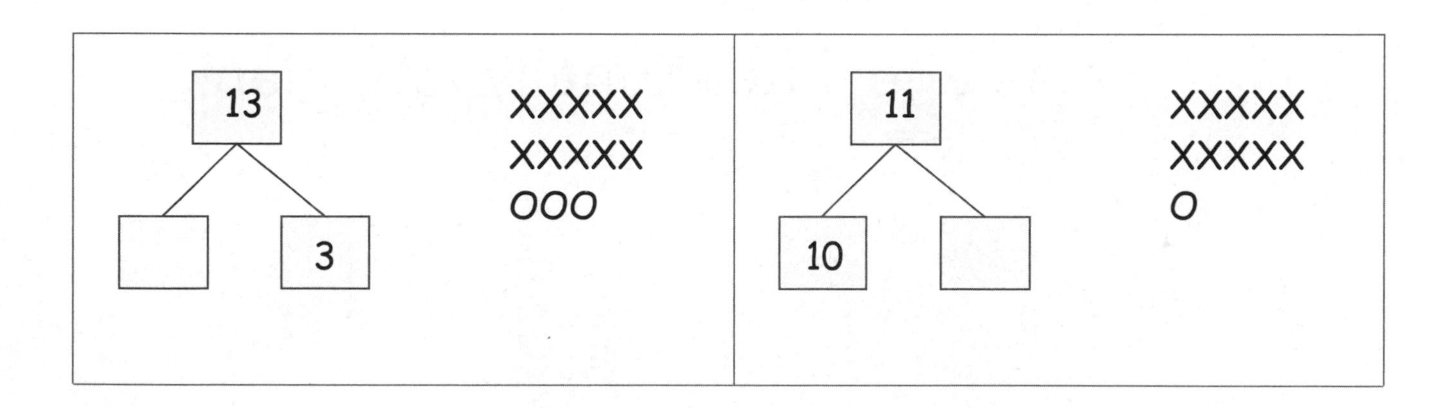

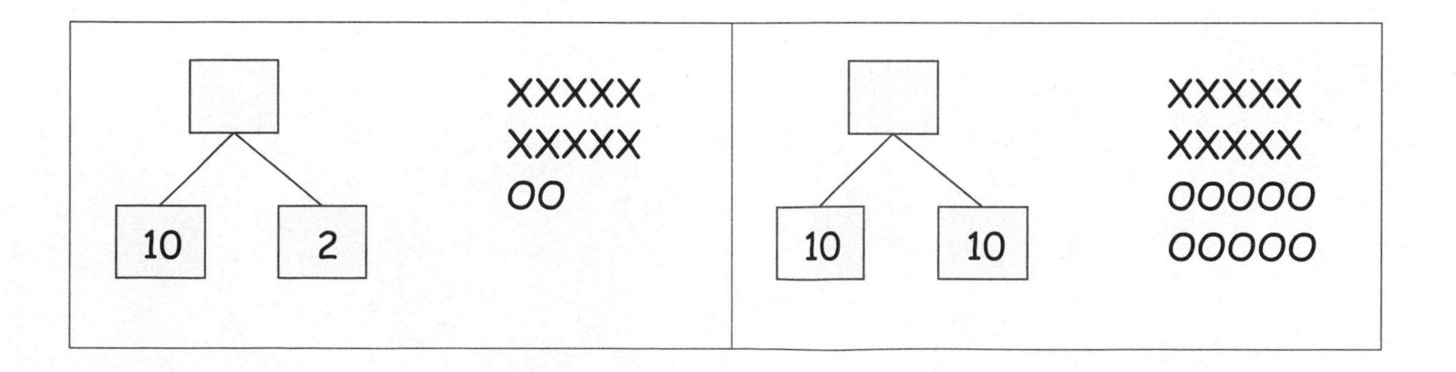

This page intentionally left blank

Name ______________________ Date ____________

Robin sees 5 apples in a bag and 10 apples in a bowl. Draw a picture to show how many apples there are.

Write a number bond and an addition sentence to match your picture.

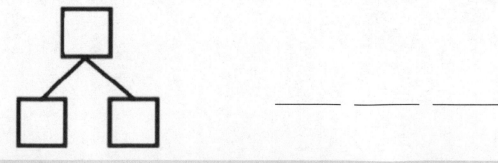

Sam has 13 toy trucks. Draw and show the trucks as 10 ones and some ones.

Write a number bond and an addition sentence to match your picture.

Our class has 16 bags of popcorn. Draw and show the popcorn bags as 10 ones and some ones.

Write a number bond and an addition sentence to match your picture.

Name ______________________ Date ____________

Bob bought 7 sprinkle donuts and 10 chocolate donuts. Draw and show all of Bob's donuts.

Write an addition sentence to match your drawing.

_______ _______ _______

Fill in the number bond to match your sentence.

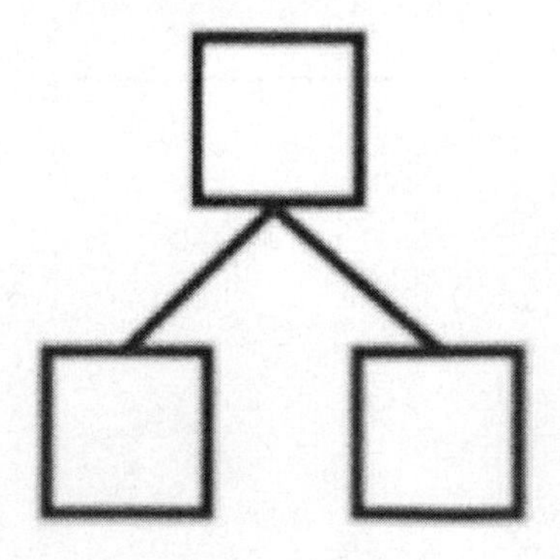

Fran has 17 baseball cards. Show Fran's baseball cards as 10 ones and some ones.

Write an addition sentence and a number bond that tell about the baseball cards.

_______ _______ _______

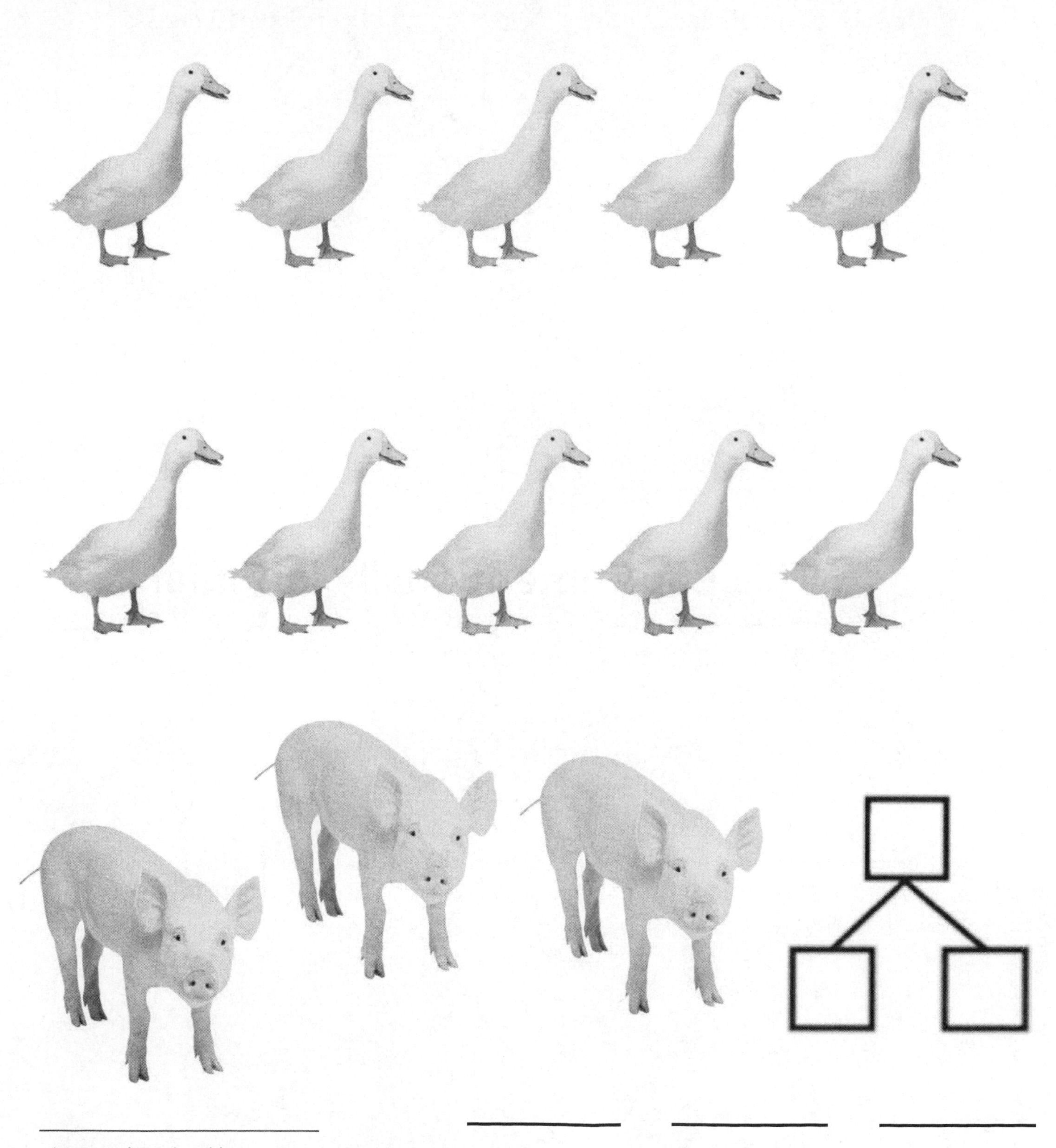

picture and word problem

This page intentionally left blank

Rabbit and Froggy's Matching Race

Directions: Play Rabbit and Froggy's Matching Race with a friend, relative, or parent to help your animal reach its food first! The first animal to reach the food wins.

- Put your Teen number and Dot cards face down in rows with Teen numbers in one row and Dot cards in another row.
- Flip to find 2 cards that match.
  Place cards back in the same place if they don't match. → 13
  Continue until you find a match.
- Write a number bond to match. → 10, 3, 13 → Hop 1 space if you get it right!
- Write a number sentence . → $13 = 10 + 3$ → Hop 1 space again if you get it right!

10	11	12	13	14	15	16	17	18	19	20

This page intentionally left blank

Student Edition

Eureka Math

Grade K
Module 6

Special thanks go to the Gordon A. Cain Center and to the Department of Mathematics at Louisiana State University for their support in the development of *Eureka Math*.

Published by the non-profit Great Minds

Printed in the U.S.A.
This book may be purchased from the publisher at eureka-math.org
10 9 8 7 6 5 4 3 2

ISBN 978-1-63255-286-0

Name ______________________________ Date ______________

Listen to the directions.

First, draw the missing line to finish the triangle using a ruler. **Second**, color the corners red. **Third**, draw another triangle.

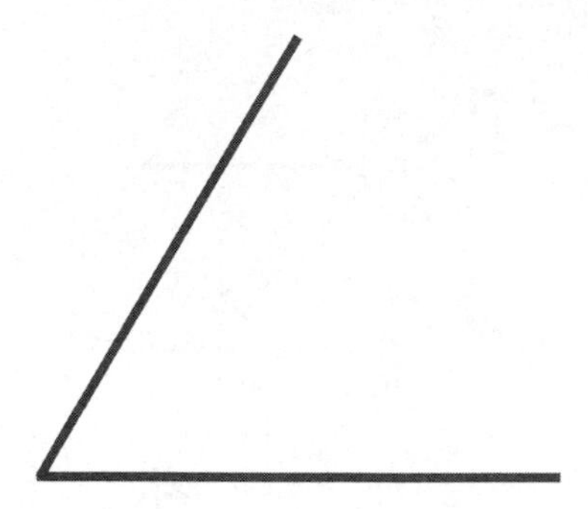

First, use your ruler to draw 2 lines to make a square. **Second**, color the corners red. **Third**, draw another square.

First, draw a triangle using your ruler. **Second,** draw a different triangle using your ruler. **Third**, show your pictures to your partner.

4 + 1 = ____

____ = 2 + 1

3 + 2 = ____

3 + 1 = ____

____ = 5 + 0

5 - 1 = ____

____ = 4 - 1

3 - 2 = ____

3 - 0 = ____

____ = 5 - 4

2 - 1 = ____

____ = 3 - 3

1 - 0 = ____

3 - 0 = ____

____ = 4 - 4

2 + 2 = ____

____ = 5 - 3

1 + 1 = ____

4 - 0 = ____

____ = 4 + 1

Name ______________________________ Date ______________

Follow the directions.

First, use your ruler to draw a line finishing the triangle.

Second, color the triangle green.

Third, use your ruler to draw a bigger triangle next to the green triangle.

First, draw 2 lines to make a rectangle.

Second, circle all the corners in red.

Third, put an X on the longer sides.

First, draw a line to complete the hexagon.

Second, color the hexagon blue.

Third, write the number of sides the hexagon has in the box below.

On the back of your paper, draw:

- A closed shape with 3 straight sides.
- A closed shape with 4 straight sides.
- A closed shape with 6 straight sides.

This page intentionally left blank

Name ______________________ Date ____________

First, use a ruler to trace the shapes. Second, follow the directions in each box. Use your ruler to draw the shapes.

Draw 3 different triangles.

Draw 2 different rectangles.

Draw 1 hexagon.

$5 - 4 =$ ____

$5 - 3 =$ ____

$5 - 2 =$ ____

$5 - 1 =$ ____

$5 - 0 =$ ____

$0 + 1 =$ ____

$1 + 1 =$ ____

$2 + 1 =$ ____

$3 + 1 =$ ____

$4 + 1 =$ ____

$4 - 2 =$ ____

$2 - 1 =$ ____

$3 - 2 =$ ____

$3 - 1 =$ ____

$5 - 0 =$ ____

$4 - 3 =$ ____

$2 + 1 =$ ____

$3 + 2 =$ ____

$4 - 1 =$ ____

$5 - 4 =$ ____

Name ______________________________ Date ______________

Trace the shapes. Then, use a ruler to draw similar shapes, on your own, in the large rectangle. Draw more on the back of your paper if you would like!

This page intentionally left blank

Name ______________________________ Date ______________

Trace the circles and rectangle. Cut out the shape. Fold and tape to create a cylinder.

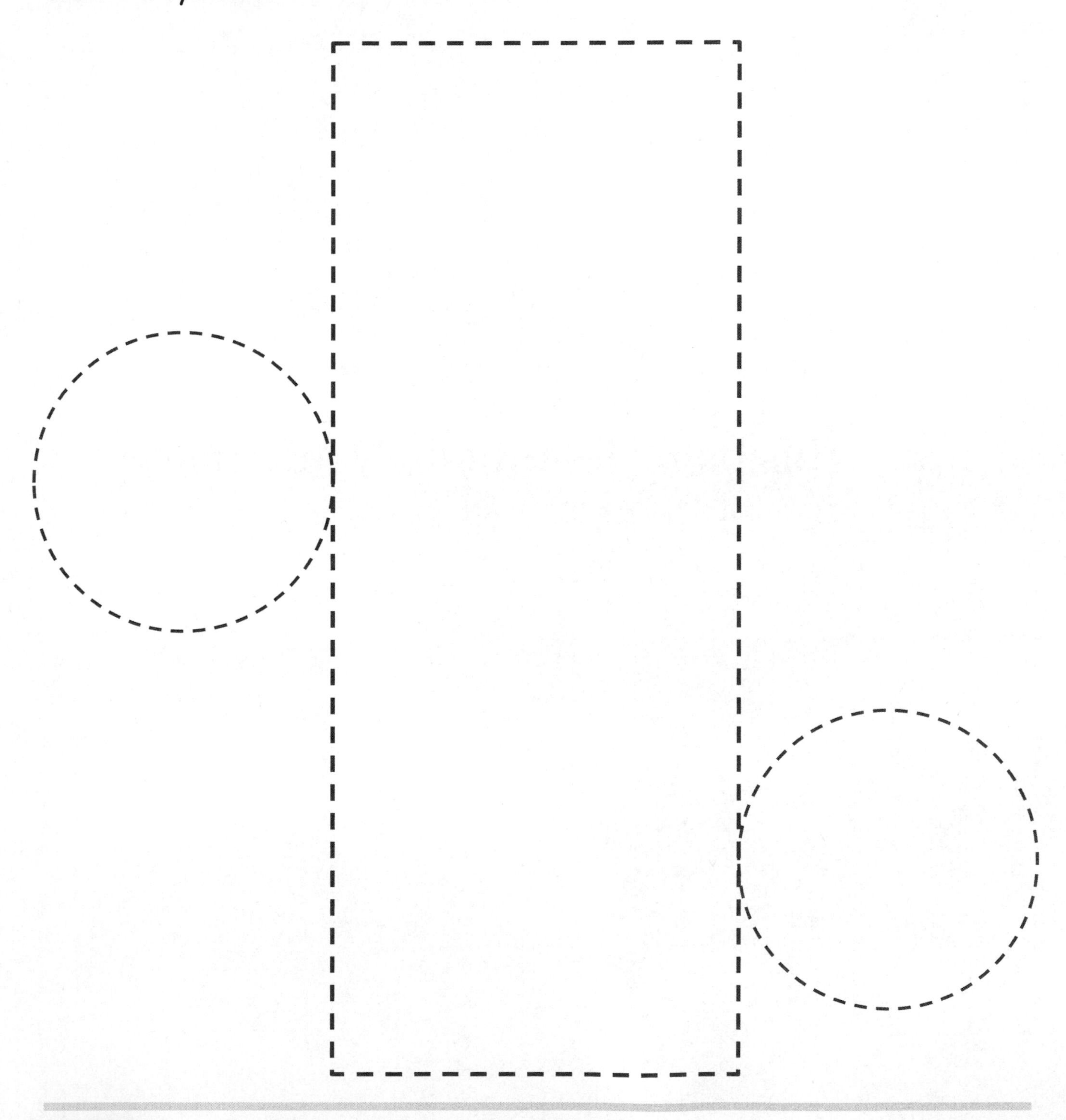

This page intentionally left blank

Trace the squares. Cut out the shape. Fold and tape to create a cube.

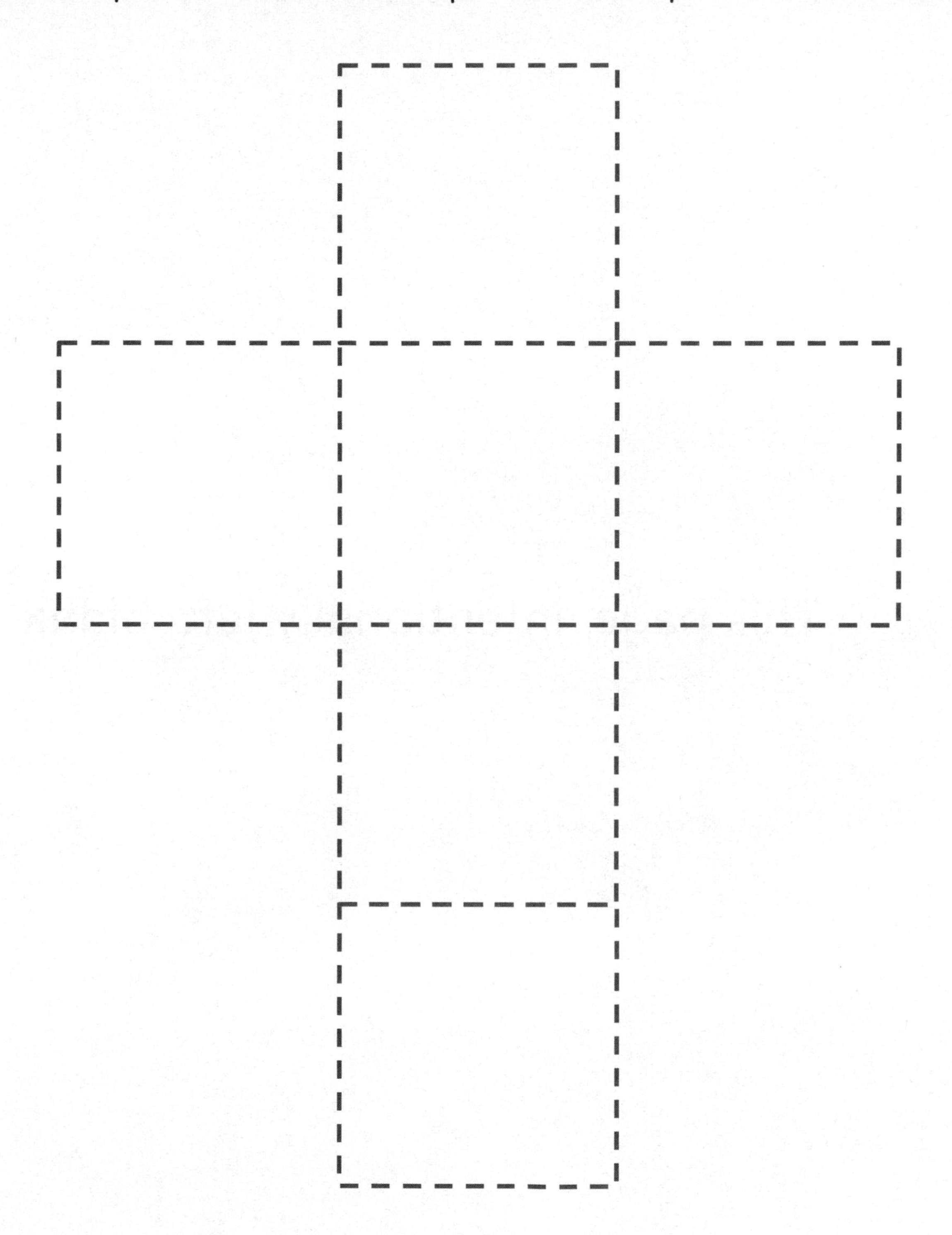

This page intentionally left blank

Name ______________________________ Date ______________

Draw something that is a cylinder.

Circle the flat shape you can see in a cylinder.

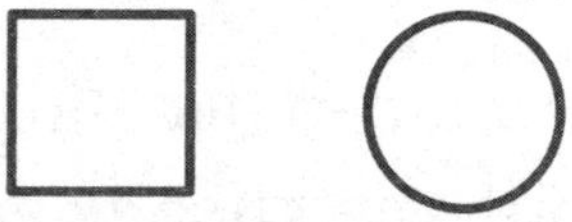

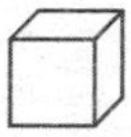

Draw something that is a cube.

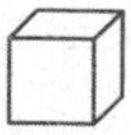

Circle the flat shape you can see in a cube.

Draw something that is a cone.

Circle the flat shape you can see in a .

Draw a 3-dimensional solid. Draw one of your solid's faces. Tell an adult about the shapes you drew.

Note to Family Helpers: Your child knows how to name some 3-dimensional solids: cylinders, cones, cubes, and spheres. You can often find these 3-D shapes around the house in objects such as soup cans, ice cream cones, boxes, and balls. For the last question, it is acceptable for your student to find and draw a different type of 3-D solid. Talk about the number of edges, corners, and faces on the object.

Name ______________________________ Date ______________

Circle the 2nd truck from the stop sign. Draw a square around the 5th truck. Draw an X on the 9th truck.

Draw a triangle around the 4th vehicle from the stop sign. Draw a circle around the 1st vehicle. Draw a square around the 6th vehicle.

Put an X on the 10th horse from the stop sign. Draw a triangle around the 7th horse. Draw a circle around the 3rd horse. Draw a square around the 8th horse.

Draw a line from the shape to the correct ordinal number, starting at the top.

9th
ninth

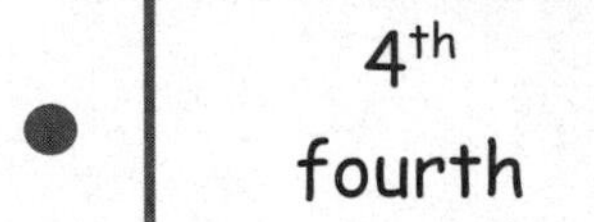

4th
fourth

6th
sixth

1st
first

7th
seventh

3rd
third

10th
tenth

5th
fifth

8th
eighth

2nd
second

Name ________________________ Date ______________

Color the 1st ☆ red.
Color the 3rd ☆ blue.
Color the 5th ☆ green.
Color the 8th ☆ purple.

Put an X on the 2nd shape.
Draw a triangle in the 4th shape.
Draw a circle around the 6th shape.
Draw a square in the 9th shape.

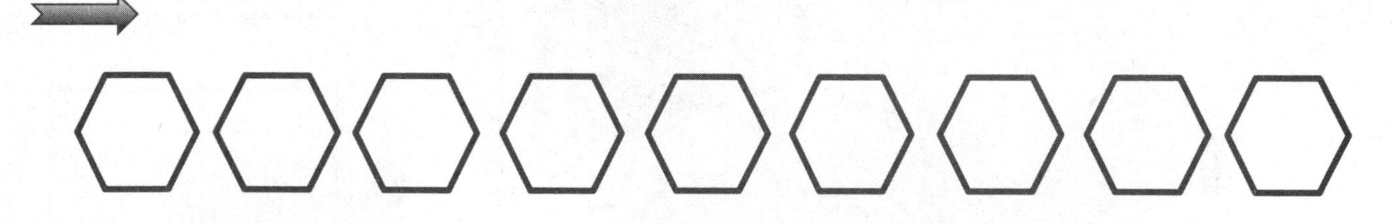

Draw a circle in the 7th shape.
Put an X on the 1st shape.
Draw a square in the 5th shape.
Draw a triangle in the 3rd shape.

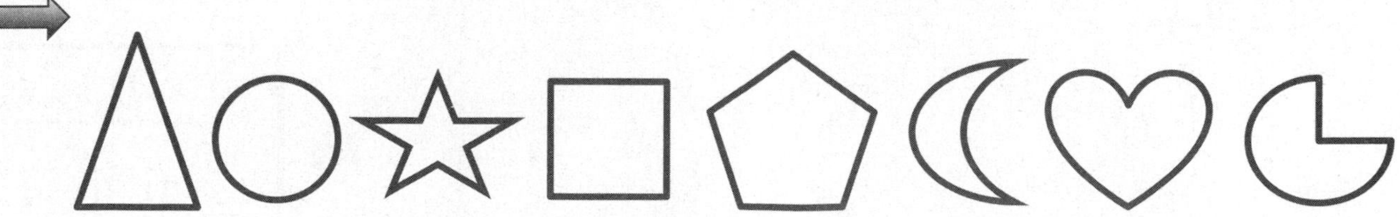

Match each animal to the place where it finished the race.

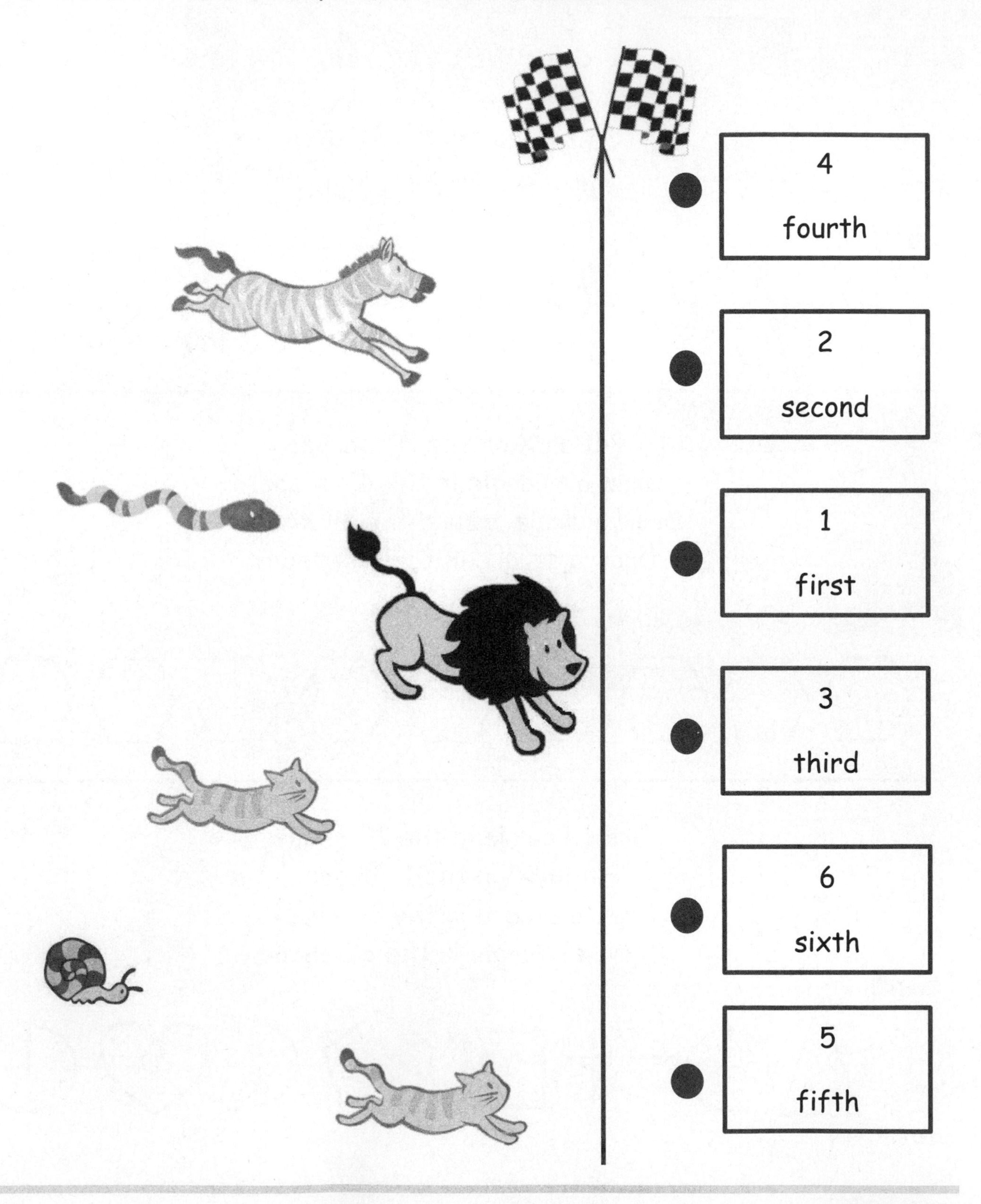

shapes

This page intentionally left blank

Name ______________________________ Date ______________

Choose 4 shapes to create a new shape in Box 1. Give the same 4 shapes to your partner. Have your partner create a different shape in Box 2.

1

2

Choose 5 shapes to create a new shape in Box 3. Give the same 5 shapes to your partner. Have your partner create a different shape in Box 4.

3

4

Subtract.

5 - 1 = ☐ 5 - 2 = ☐ 5 - 3 = ☐ 5 - 4 = ☐

Name ______________________ Date ____________

Match each group of shapes on the left with the new shape they make when they are put together.

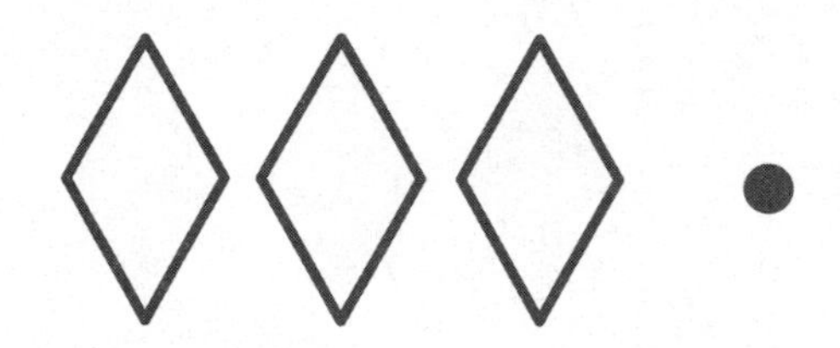

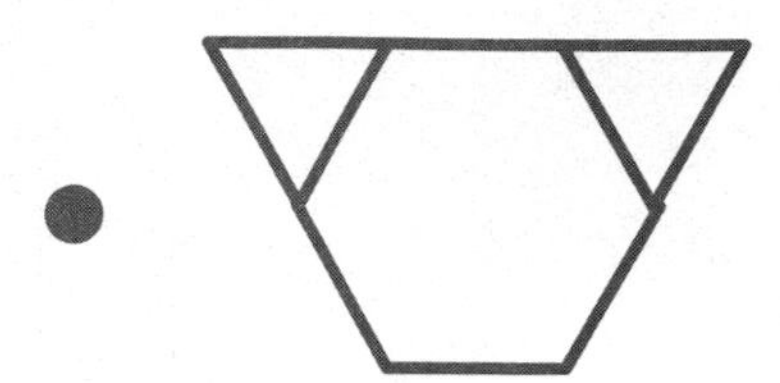

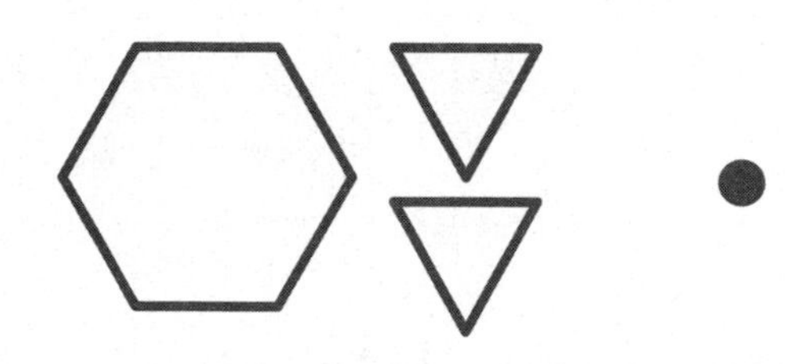

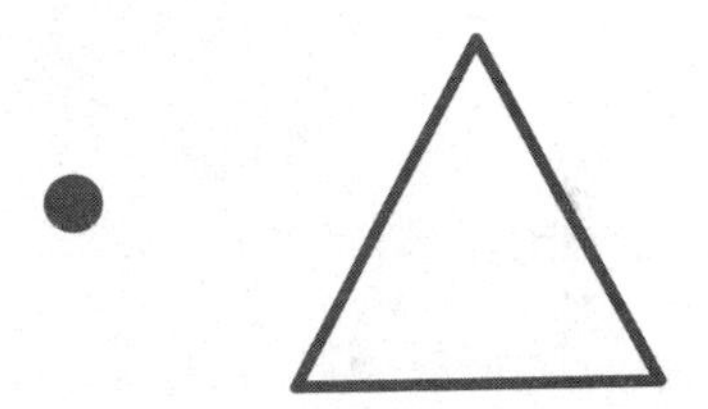

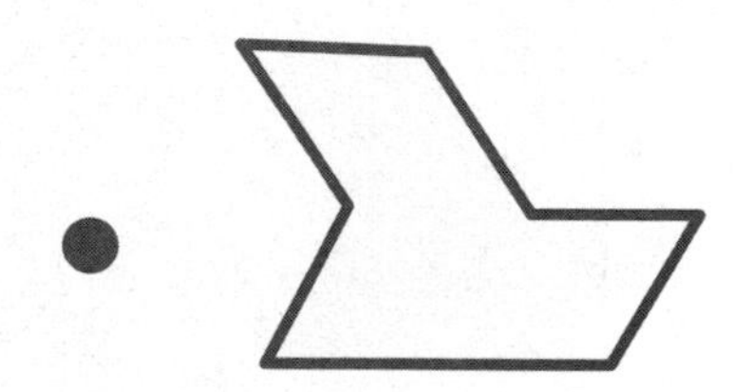

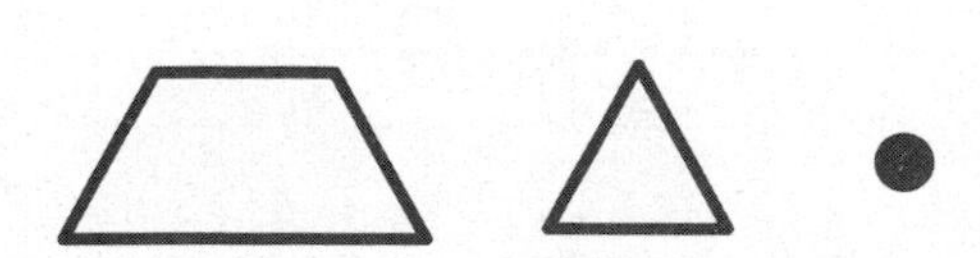

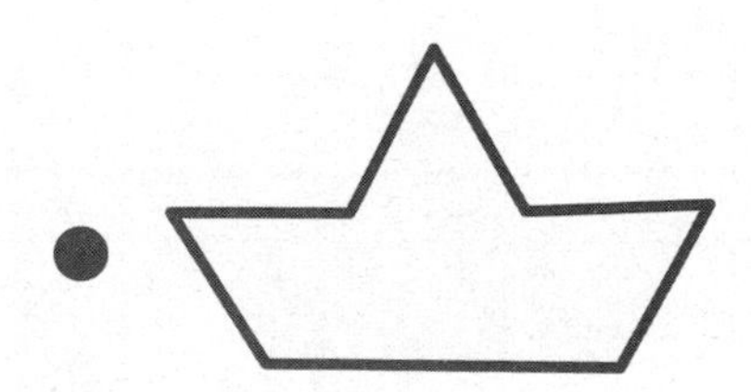

This page intentionally left blank

I Can Make New Shapes!

I can make new shapes recording sheet

This page intentionally left blank

Name ______________________________ Date ______________

Trace to show 2 ways to make each shape. How many shapes did you use?

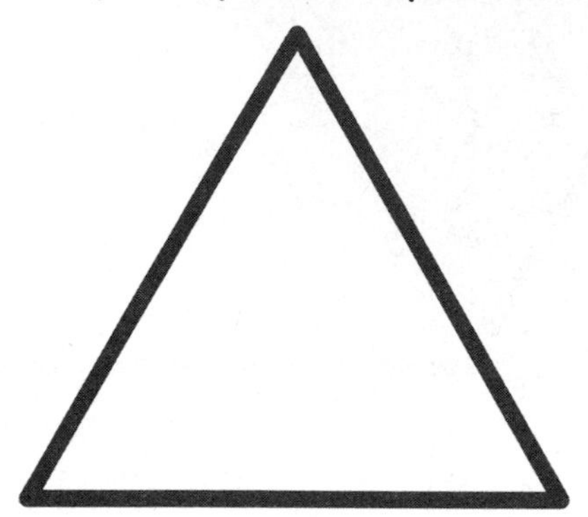

I used 3 shapes.

I used ____ shapes.

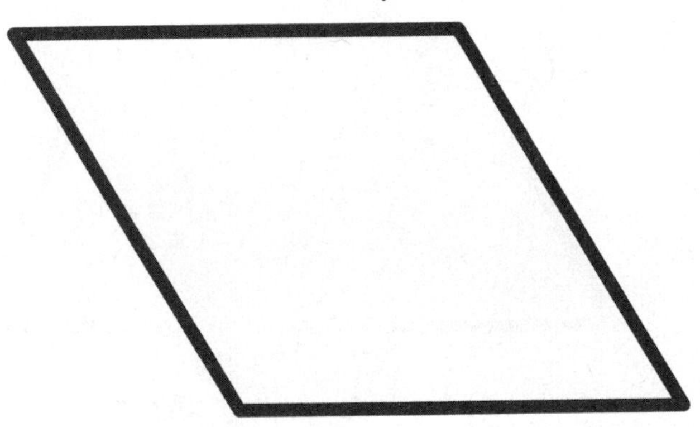

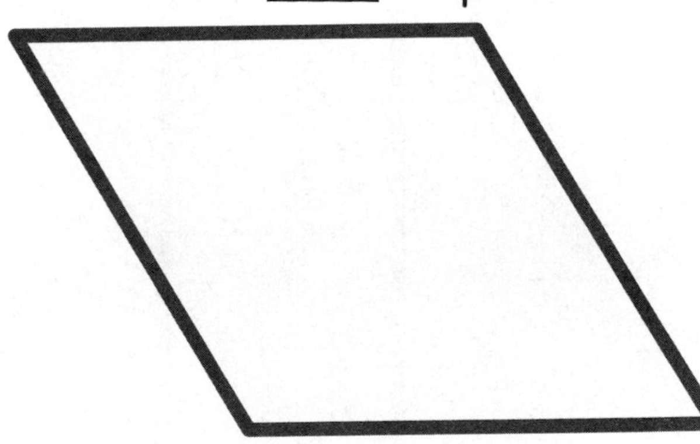

I used ____ shapes.

I used ____ shapes.

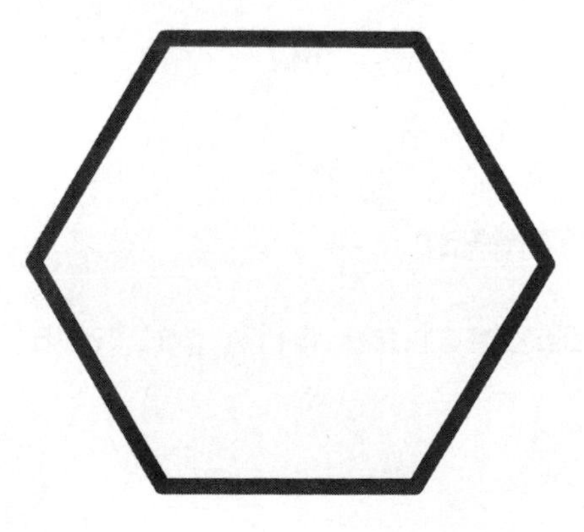

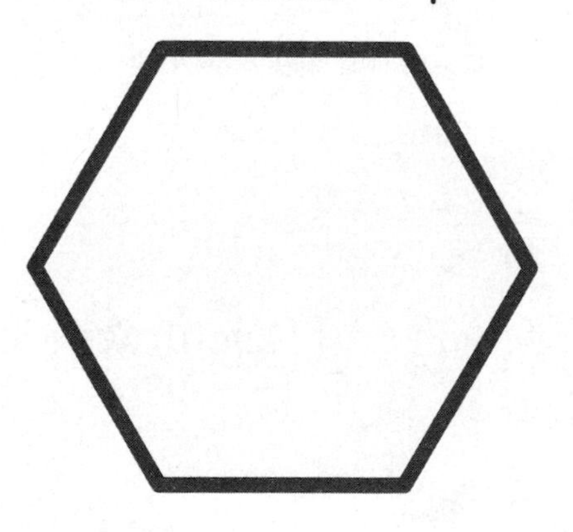

I used ____ shapes.

I used ____ shapes.

Fill in each shape with pattern blocks. Trace to show the shapes you used.

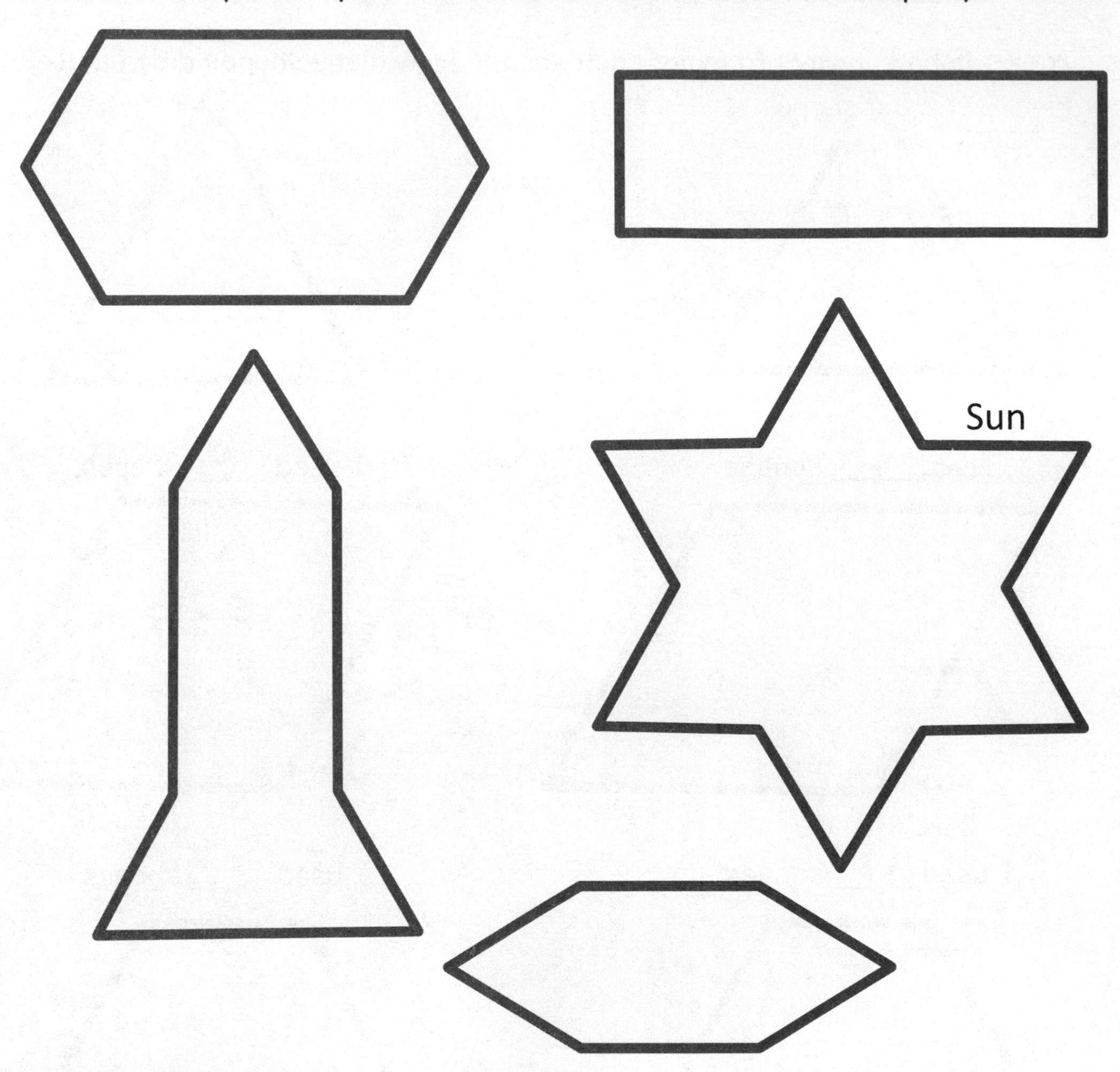

How many different ways can you cover the sun picture with pattern blocks?

Name ______________________________ Date ______________

Cut out the triangles at the bottom of the paper. Use the small triangles to make the big shapes. Draw lines to show where the triangles fit. Count how many small triangles you used to make the big shapes.

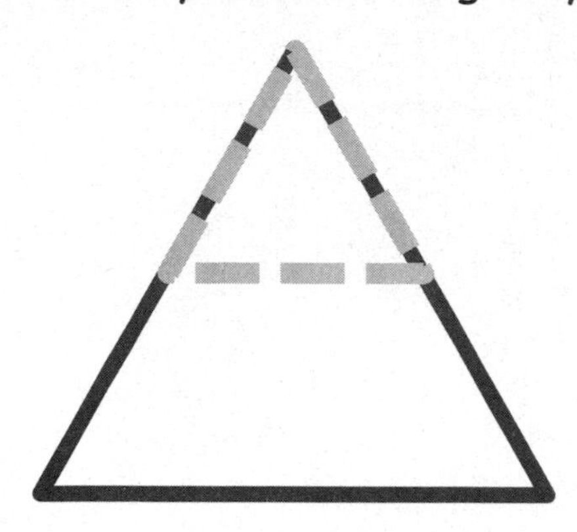

This big triangle is made with ____ small triangles.

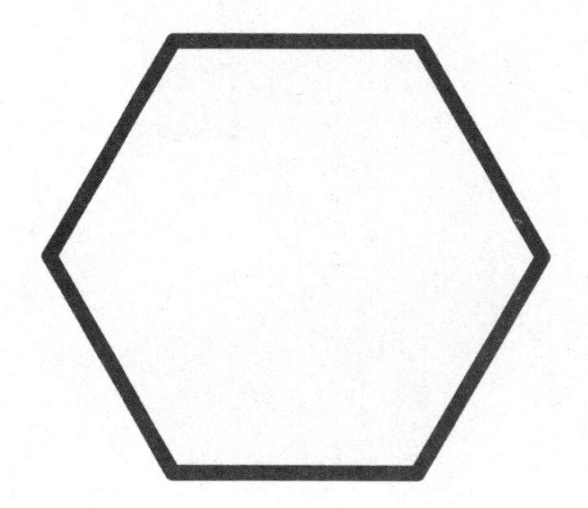

This hexagon is made with ____ small triangles.

This page intentionally left blank

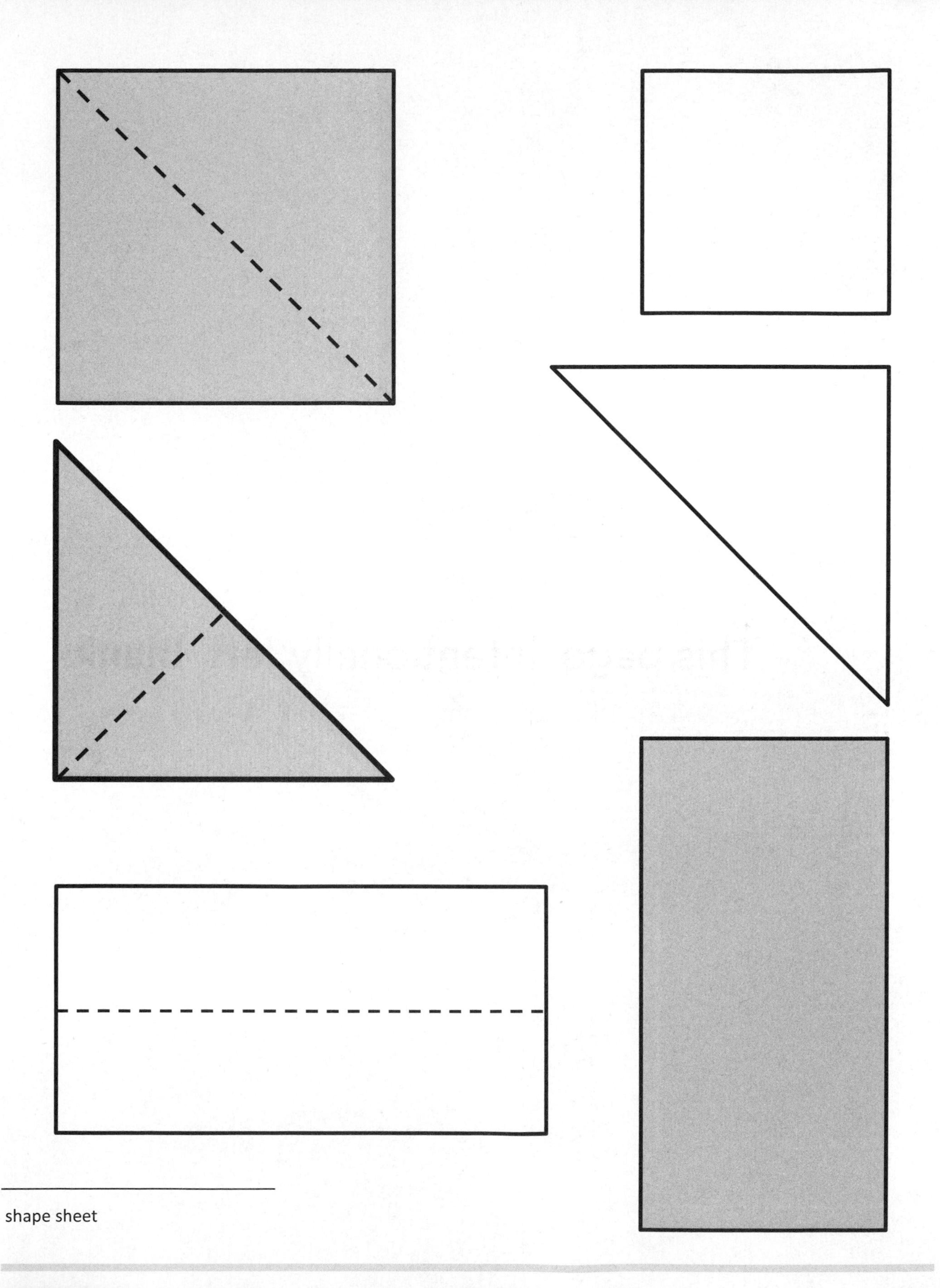

shape sheet

This page intentionally left blank

Name ____________________ Date __________

Glue your puzzles into the frames.

Glue puzzle here.

Glue puzzle here.

Draw some of the shapes that you had after you cut your rectangles.

Carlos drew 2 lines on his square. You can see his square before he cut it. Circle the shapes Carlos had after he cut.

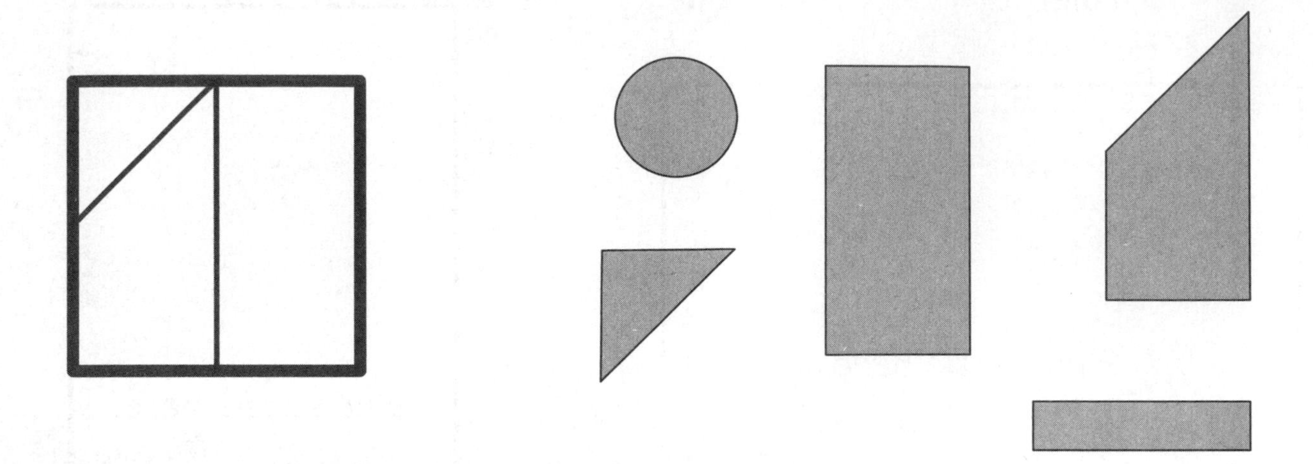

India drew 2 lines on her rectangle. You can see her rectangle before she cut it. Circle the shapes India had after she cut.

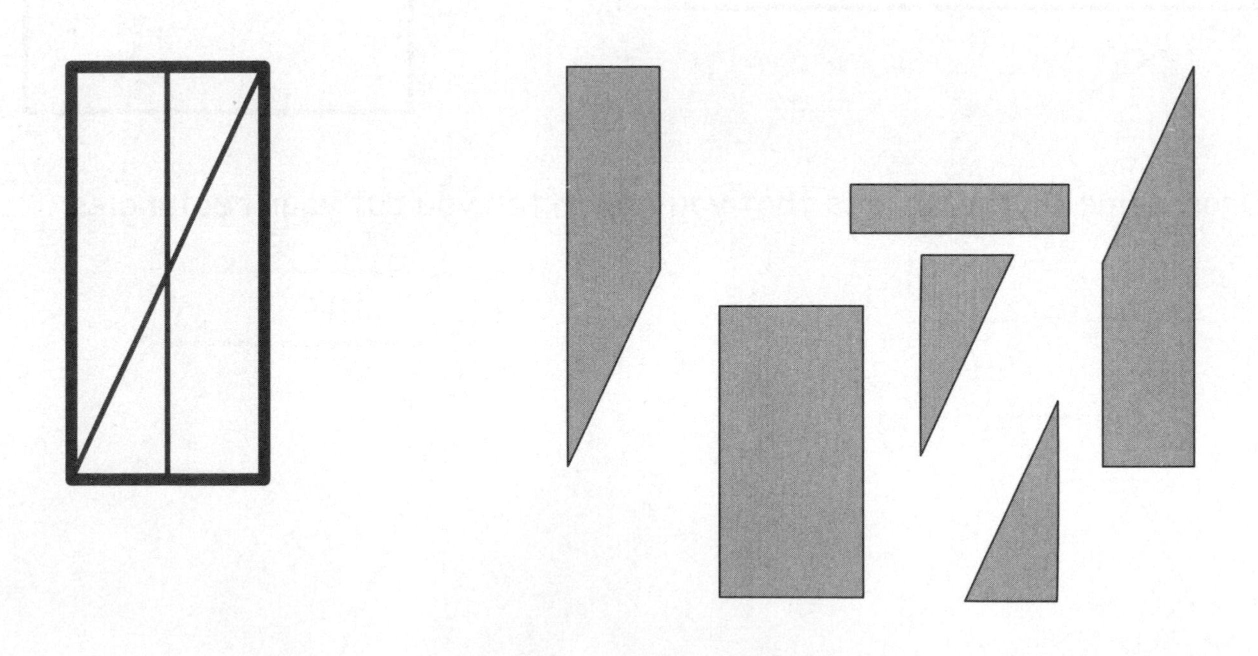

Name ______________________________ Date ______________

Using your ruler, draw 2 straight lines from side to side through each shape. The first one has been started for you. Describe to an adult the new shapes you made.

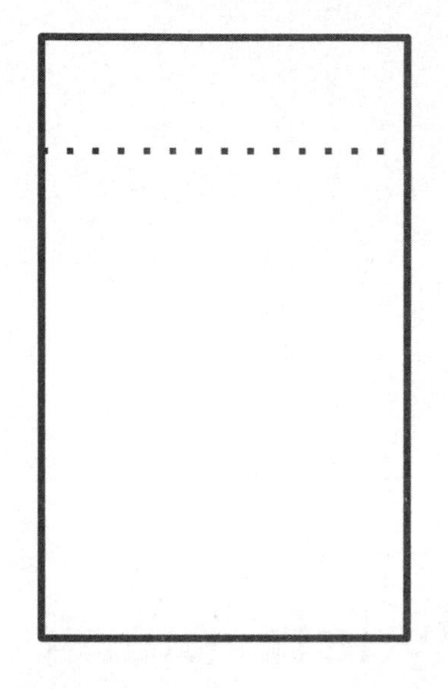

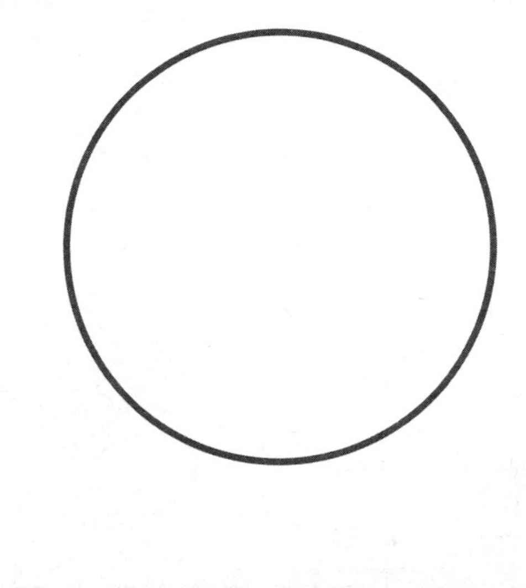

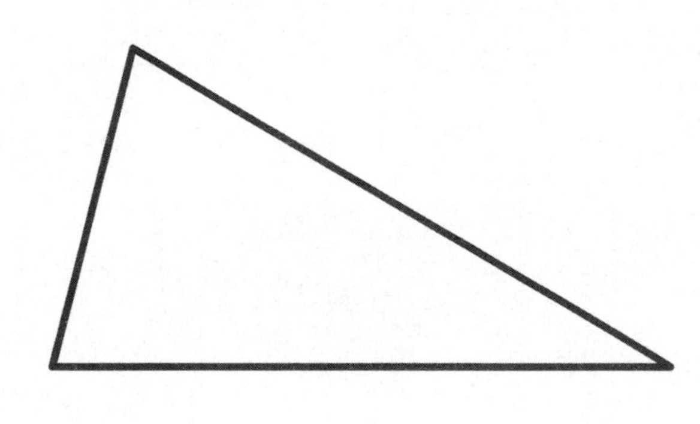

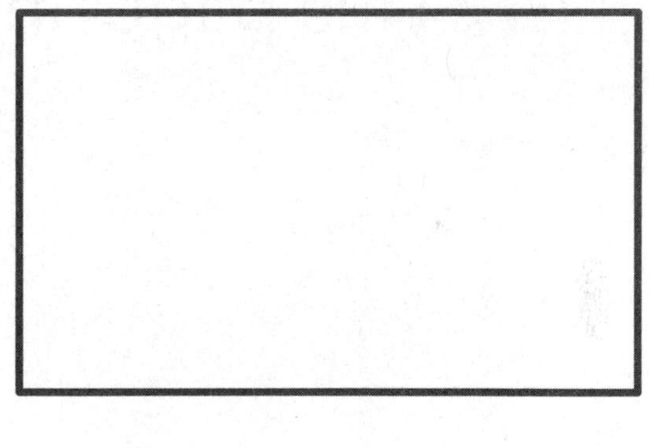

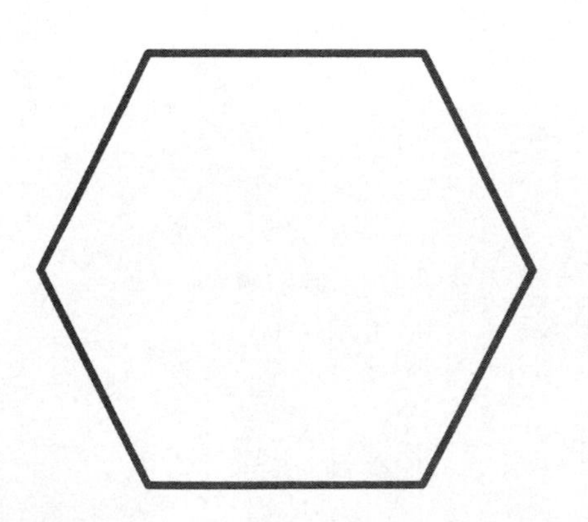

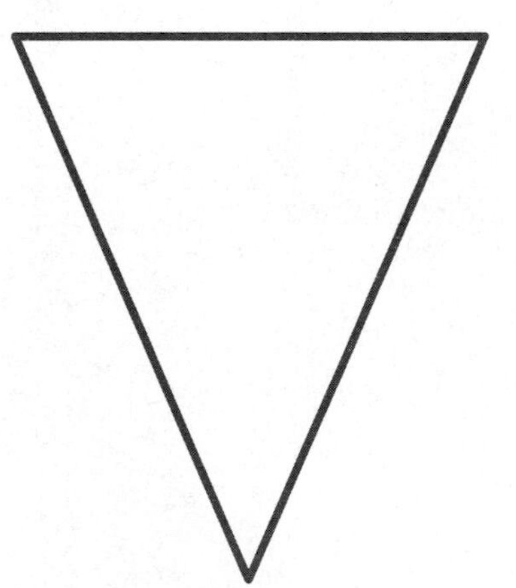

This page intentionally left blank

shape puzzle

This page intentionally left blank

Name ______________________________ Date ______________

A. Make-10 Mania: Show how you made 10.

Name ______________________________ Date ______________

B. Five-Group Frenzy: Write the number, draw the number in the 5-group way, and draw the number in any other configuration.

This page intentionally left blank

Name ______________________________ Date ______________

C. Shape Shifters: Choose 5 pattern blocks, and create a shape. Trace your shape, and then trade with a partner.

Name ______________________________ Date ______________

D. The Weigh Station: Choose an object. Guess how many pennies are the same weight as the object. Then, see if you guessed correctly! Draw a picture of the object, and write how many pennies it weighs.

This page intentionally left blank

Name ______________________________ Date ______________

E. **Awesome Authors:** Roll the die. Use the number to create an addition or take-away sentence. Draw a picture, number bond, and number sentence. Share your story with a friend.

- -